CW00467855

The Digital Workplace

How Technology is Liberating Work

The Digital Workplace

How Technology is Liberating Work

Paul Miller

TECL Publishing
New York – London

Copyright © 2012 by Paul Miller

All rights reserved. No part of this publication may be reproduced, distributed, or transmitted in any form or by any means, including photocopying, recording, or other electronic or mechanical methods, without the prior written permission of the publisher, except in the case of brief quotations embodied in critical reviews and certain other non-commercial uses permitted by copyright law. For permission requests, write to the publisher, addressed 'Attention: Permissions Coordinator,' at the address above.

All trademarks are the property of their respective companies.

Published in the United States by Dog Ear Publishing LLC
4010 W. 86th Street, Suite H, Indianapolis, Indiana, 46268

Miller, Paul
The Digital Workplace/How Technology is Liberating Work/ Paul Miller – 1st edition.

ISBN 978-145751-096-0
1. Business and Management. 2. Technology.

Design and typeset by:
Toast Design Consultancy Ltd, 16 North Bar St, Banbury, OX16 9AT
Tel: + 44 (0)1295 266644

First Edition

About the author

Paul Miller is a technology and social entrepreneur. He is the CEO and Founder of the Intranet Benchmarking Forum and the Digital Workplace Forum, host of IBF Live, a monthly intranet media show, and hosts IBF 24, an annual online 24-hour workplace technology event.

After an early career as a business editor and speechwriter, Paul founded the influential *WAVE* magazine in 1990, a forerunner to *Fast Company* and *Wired*. He co-founded The Empowerment Group in 1992, pioneering new approaches to communication within major organizations. In 1993, he co-founded the Ideas Café, a regular innovation gathering shaped along social software lines during the early days of the web. He wrote the knowledge management handbook *Mobilising The Power Of What You Know* (Random House). He lives in London and has two daughters.

Intranet Benchmarking Forum

Established in 2002, the Intranet Benchmarking Forum is the world leader in intranet benchmarking, research and best practice. Its members are all Fortune 500 and equivalent organizations. Around 20% of the world's top corporations are members of this confidential community, which has shaped the intranet industry globally.

Digital Workplace Forum

The Digital Workplace Forum is the world leader in digital workplace measurement, connection and best practice. Established in 2011, its members are all Global 500 organizations committed to defining, investigating and shaping the new digital world of work.

Contents

Introduction

Work is one of the persistent aspects of being human. Food, child-rearing, shelter... and work... have been some of the enduring characteristics of our species. Irrespective of what the next centuries and millennia may bring, it is impossible to imagine human beings who never work. In *Star Trek*, physical travel was replaced by instant teleportation, but the crew were still workers. Work matters – it always has and always will. So, both *where* and *how* we work deserve our close attention, since they crucially affect how we live, who we are and what we become.

In 1794 the first incarnation of the telegraph was invented by Claude Chappe[1] and, in 1876, Alexander Graham Bell took this technology a stage further with the invention of the telephone. Both devices were used in the workplaces of the 19th century and brought with them a non-visible, non-physical, let's say *virtual*, working space. It must have been beyond comprehension when, on 9 October 1876, Bell and Thomas A Watson held the first wire conversation ever and they talked to each other by telephone over a three-kilometre wire stretched between Cambridge and Boston.[2]

Work has historically been about location and tools – whether it be agricultural, manufacturing, services – and we have traditionally worked where the necessary premises, facilities and tools were located. We lived where we worked or travelled to places where the machinery and people we required could be found. Communities, villages, towns and whole cities grew up around the Physical Workplace.

In the last two centuries, these working environments have been transformed from the brutal, harsh workhouses chronicled so vividly by Charles Dickens,[3] to the only slightly less painful factories and dreary, monotone offices of the last century, to the hip, funky spaces which house modern workers today – awash with cafés, internal malls, play rooms and airy, comfortable meeting spaces. Since we needed to be physically located somewhere in order to achieve a day's work, employers gradually became smart enough to realize that making the Physical Workplace attractive and a pleasant place to be in would produce a better, more effective workforce.

But the journey which began with that phone call in 1876 has been relentless, albeit slow for the most part due to the paucity of 'game-changing' technical innovation. While the physical spaces of work have been evolving before our eyes, the 'non-physical spaces' have also been shifting, but without this being fully recognized because the impact has been only modest in the past 100 years. However, in the last 20 years, and particularly the last five years and specifically the last two years, the progress has accelerated wildly, bewilderingly, as a 'perfect storm' of technological innovation, universal access, cheap (almost zero) costs has erupted around us. It is as if the 100 years and more of pedestrian growth has exploded with a volcanic force, shifting and rupturing everything we touch as we work.

We tend to notice what we can see with our eyes. But The Digital Workplace – this world that merges work and technology – is invisible for the most part and definitely can't be grasped physically like a chair or a desk or a pen. If we could actually see the current digital work world encircling us, it would leave us breathless, given its reach, depth and moving parts; it would be not just a new continent but a new planet (only digital). Predicting futures is always imprecise but what we can say for sure is that the trajectory of Physical Workplaces will continue to be interesting but unremarkable. Yes, offices, factories and retail outlets will be redesigned to suit new conditions, but the changes will not be transformational because each physical place will become less and less central to work itself.

If we could actually see the current digital work world encircling us, it would leave us breathless, given its reach, depth and moving parts; it would be not just a new continent but a new planet (only digital).

What *will* be transformational will be the new geography of work: the Digital Workplaces where we will spend more and more time, working in entirely new ways, with richer, more immersive tools. This new digital working world is always there, always on. Whether we are in our traditional or newly repurposed offices, or wherever else we happen to be working; at home, in the garden, in a co-working space, a café, on a train, in the park, in a car... My focus in this book is on the *digital* in the workplace as we start to navigate a new, as yet unexplored, world of work.

At the moment we are still taking our first baby steps in the Digital Workplace but in the coming years and decades ahead we will get to know this new terrain very well. We will need to influence the map so that this new world of work suits us and supports us. We will develop new skills, behaviour and language as we fly across time zones and regions and between people. It will be deeply challenging and yet very exhilarating.

As I have found through my 30-year working life in various incarnations of the Digital Workplace, an unexpected and yet beautiful aspect of the digital working world is that, while Physical Workplaces restrict, almost *imprison*, the workforce, the Digital Workplace has the opposite effect. It liberates us.

MY
WORLD

Chapter 1

So, is this it?

Most work in most organizations is awful. It's mind-numbing, soul-destroying and depressing. Everyone knows this but virtually no one admits it. Walk into most offices and you can feel your heart sink. Same place, same tedium, just another day. The supposed 'best' jobs are just as bad as the so-called 'worse jobs' – bigger desk, bigger office, bigger salary – but the daily drudge is just the same.

Study hard, pass your exams, get a good degree and you too can land the job of your dreams, only to discover... you have arrived at nowhere. Most people are dead on their feet before they even start work each day. They've dragged their bodies to the workplace by train, bus or car, and battled through the chaos of the dehumanizing morning commute – all in order to find themselves back between the very same four walls they were so glad to vacate the evening before. Welcome to the working world!

"Without work, all life goes rotten. But when work is soulless, life stifles and dies."

Albert Camus

For better or for worse, I realized this fundamental truth about work on my very first day in my first proper job. It was 16 August 1979 and I was starting work as a trainee reporter on the *Newcastle Evening Chronicle* in the north-east of England. I was 21 years old and had just graduated in law from Birmingham University. I hated law but had been passionately interested and active as a writer and ad hoc journalist throughout my time at university, having written for university publications and local newspapers. Out of 3000 applicants for the trainee reporter job, 20 had been selected for the Thomson Newspapers Graduate Training Scheme and two of the 20 were to start on the *Chronicle*. I was one of those two. It was a dream job and had taken a lot of effort and good fortune to be selected.

So, at 8.30 sharp, I arrived at the *Chronicle* offices and was shown to my desk in the newsroom; it was buzzy, exciting and real. The hard-nosed News Editor gave me a press release to 'turn into copy' and I set to work on my typewriter (it was a long time ago!). Midway through the morning a gnawing sense of anxiety arose in my stomach: 'Is this it?' I wondered. 'This is the exactly the job I wanted, longed for… but do I really have to just sit here in this stuffy room and do what someone tells me all day?'

Later that evening, my landlady asked how Day One had gone. 'Not bad,' I replied. 'But there's something bugging me.' 'What's that?' she asked. 'Well, it's just so odd that someone pays you money and that means you have to sit at a desk all day. It's like you've sold your soul,' I said. 'Don't worry,' she replied. 'You'll get used to it.' But I never did.

That conversation encapsulated what I thought at that time and still think (more than 30 years later) about the nature of work. I never did get used to that sense of having sold something you never should sell. And the landlady had unintentionally identified the flipside of the situation: that most work is, in varying degrees, depressing but you just acclimatize to it. After a while, you cease to notice what you've settled for, justifying it with the argument that you have to work in order to earn money.

For more than three decades the 'revelation of 16 August 1979' has colored my attitude to work and, in the main, I have really enjoyed my working life. After a few years of employment in supposedly 'sought-after' jobs in various countries, followed by a period of freelancing, I took the initiative to change things and embarked upon life as an entrepreneur. Since then, I have created several businesses.

Now, you could argue that I am just a maverick (with some justification) and an entrepreneur by nature and therefore unsuited to working conformity, but, while this may be to some extent true, I also believe that my insight into the real nature and value of most work is borne out time and again, day in day out, among people I meet. Either because they are still laboring in roles they basically dislike, or because they have managed to escape into newly liberated work situations that allow them to look back and reflect on just how unsatisfactory their former work situation was.

Imagine this

What I didn't perhaps understand immediately, but can see now, is just how *physical* the problem is. Imagine a different start to my working life that goes like this… *Even before Day One, I can log on to a secure 'new hires' section of the newspaper's intranet and connect with new colleagues… see what people are reporting on today… watch the web cameras trained on the newsroom floor. Then, on arriving in the actual office, I am presented with fully mobile tools and introduced to a range of technologies that will enable me to work more or less when I want and from where I want… 'So, Paul, you can either work here in the office – but do pre-book a desk if you are going to do that – or from home, or from a café with Internet access, or from one of the co-working hubs around the city where some of the guys meet up. As your editor, I need to be in contact with you easily and frequently throughout the day via mobile, instant messenger, teleconferences and sometimes in person, but all stories are part of a workflow on the intranet. This is not the easy life though, Paul. You need to deliver accurate, strong stories quickly, on time and regularly… how you achieve that is pretty much up to you.'*

If Day One had been like this, then what would my landlady and I have talked about that first evening, I wonder? My disquiet, I now realize, was not with the job itself, which was tremendous fun and frequently exciting, but with the intrinsic design, shape and pattern of the daily work.

Gentrified 'slavery'

The issue of physical control is still fundamental to work in most organizations today. Looked at objectively, I believe modern work shares some of the deeply unpleasant characteristics of a historical form of slavery, albeit in a 'gentrified' form. Work often involves control over the physical movement of staff: 'I want my people here at their desks each day, all day' say many managers – just as slaves were, and still are, restrained physically in order to control their movement.

The other key aspect of historical slavery is that slaves were tethered, then supervised and monitored to check that they were working as required. The modern workplace equivalent is the manager who operates by only feeling secure and in control if they can directly watch their team at work at their desks. This is revealed when managers resist the notion of their staff working outside the office because 'I don't want them wasting time watching TV at home – I need them here where I can keep my eye on them.' The key concern for such managers with portable work practices is the fear of losing control. These two aspects of most work still create a gentrified type of slavery, based on physical control and direct observation.

Of course, slaves in the past received (and indeed modern examples of this persist the world over today) no pay aside from 'benefits' in the form of shelter and food. Plus, slavery held a sanction through violence, so this comparison has its limits but, still, too many managers display no trust in those who report to them and this surfaces when they are questioned about how the Digital Workplace might work for their team. Compare that with staff and contractors who can operate wherever they wish and whenever they want and this includes – if it works best

Looked at objectively, modern work shares some of the deeply unpleasant characteristics of a historical form of slavery, albeit in a 'gentrified' form.

for them and for their colleagues – spending their days in the traditional office. The ability to choose and flex, based on what is needed, breaks the yoke of physical restraint for the first time. Overwhelmingly, people then feel a deep sense of release and liberation, as if an invisible chain has been unlocked.

What we are seeing happening around us now, in virtually every aspect of work and across most sectors, is the liberation of the workplace tools and technologies so that they are not physically dependent on any one location.

It's not where we work but how we work

The nature and design of work are shifting – and fast. In recent years, what has begun to emerge is a new way of working based more around how we work than where we work. I have come to see that my issue with work in 1979 – and this is still the case for most work in most organizations – was about autonomy, control and power. In the 'fast-forward imaginary 16 August,' the difference is not the job itself but the way the work is shaped. The big advance in the imaginary scenario is that the individual has been granted the power to influence and control their own work, handing them the freedom to shape their working day to suit themselves within reasonable constraints, the only requirement being the responsibility to deliver successfully and on time.

When I say that the issue is *physical*, I mean just that. The Physical Workplace restricts employees to being physically in a fixed place at a fixed time – relentlessly. Until recently, there was no alternative. You had physically to go to the office because that was where the work tools were located. At the end of the working day, the work and the tools stayed there in a drawer; and then you came in again and took them back out to resume the next day… and the next… and the next…

What we are seeing happening around us now, in virtually every aspect of work and across most sectors, is the liberation of the workplace tools and technologies so that they are not physically dependent on any one location. In many jobs, in many organizations, both large and small, you can now work from anywhere – and, in so doing, you begin to move towards attaining that elixir we all crave, whether consciously or not: a sense of autonomy, flexibility and power over the working day. This need not be 100% power, but even a degree of autonomy and influence goes a long way towards overcoming the 'numbing' effect that has so often accompanied work in the past.

One recent trend that has developed since I began writing this book is that, in many countries, changing the shape of work is becoming government policy. In Holland, where there is now a 'New Ways of Working' week every November, the key government motivation is traffic reduction; in the UK, initiatives have been driven by the 2012 London Olympics, by rush-hour concerns and through cost-cutting across the public sector; in the US, the driver is once again cost reduction, focusing on unnecessary travel and redundant use of paper, printers and office equipment in the public sector.

And I'm born again

In fact what has transpired is that my experience of work after my time as a journalist has shown itself to have many of the patterns that are now taking place quite generally across the world of work, enabled by better, faster, cheaper technology.

Not being able to see an alternative in 1979, I chugged along at the *Chronicle*, working around the system as best I could to get the job done while at the same time attempting to fashion some degree of control. I would come up with 'in-depth' stories that entailed my being 'out and about' for days at a time while researching the background. Then, in 1984, after five years of various jobs and promotions that, along the way, took me to Australia before returning to London's Fleet Street, I quit to work freelance as a journalist, copywriter, speechwriter and ghostwriter.

One day, right at the start of my time as a freelancer, I began the day by planning a speech I'd been commissioned to write for a senior banker. Then I went for a long walk and took in a film at my favourite local cinema (the ideal preparation for any intense piece of writing) before sitting down – at home – to write solidly for most of the next 24 hours. As I sat there with the speech taking shape, it struck me that not only was I earning in one day the equivalent of a month's salary as an employee, but also – most importantly – I was in control; setting the agenda, the timing, the whole process. I felt alive in work for the first time.

Quite soon after this, I co-founded a medium-sized business, The Empowerment Group (TEG). This was in the 1990s and, while innovative in its work, the company essentially 'aped' the large corporate clients we serviced. TEG delivered change communication programs for global organizations and we did all the things they did – but on a smaller scale. They had offices; so we got an office. They had secretaries; so did we. They owned and ran their IT; we did too. They had as many staff as they could get away with and everyone commuted into HQ… day in, day out… as did we! For me, this was a backward step. Admittedly, I enjoyed the creative, innovative aspects of running a decent-sized consultancy but, at the same time, there I was, back commuting into an office again. Hadn't the intention been to leave that behind?

Digital from Day One

In 2002, I founded the technology research company, Intranet Benchmarking Forum (IBF), which has become the world leader in intranet measurement, best practice and community building, servicing a long list of Fortune 500 and equivalent organizations globally.

From the outset of IBF, I decided that it must be possible to retain the freedom and joy of working freelance, even within an organization. Building on experience, I unconsciously chose to 'do the right thing' – whatever that might be – rather than the 'expected' thing. From the very start, IBF was virtual, networked and flexible. We had a small office at the bottom of my back garden in north-west London but the maximum number of people who ever worked there was four.

> The more flexible you are, the better people you can recruit – and great people make great businesses.

The business grew fast and quickly developed from the handful of people involved in the early days to a current count of around 50. We have expanded from the UK into North America and mainland Europe, opening what we might term 'offices' in New York, Zurich, Sydney and Los Angeles; in reality, these are 'virtual offices' only, operated by regional IBF staff and freelancers.

Eventually, in early 2011, we ditched what we rather grandly used to call the 'Headquarters,' that is, our one small office because, increasingly, no one felt the need to go there. This move went without any glitches. We have always worked 'virtually,' partly because it made sense commercially but also because of the benefits in terms of staff morale. The more flexible you are, the better people you can recruit – and great people make great businesses.

Quite suddenly, it seems, we are beginning to witness all around us the Digital Workplace starting to take shape and to transform the nature of work. The Digital Workplace has many aspects and consequences, as we shall see in the course of this book, but partly what we are seeing is the 'freelance' mode of operation becoming the working norm in all sorts of jobs.

For instance...

We might find Sylvia, who is employed by a large financial services company, choosing to work from home this Thursday and Friday to complete some work on next year's strategy plan; on Monday she'll present it via web conference to her global team; but on Tuesday and Wednesday she'll be back in the office for meetings... and so on...

Wander into the nearest Starbucks. Of the people around you on the phone, on a laptop, chatting... which ones are employees, which freelance, who is in a business meeting with a colleague, who's just hanging out with friends, who is working, who's studying? You can take a guess, but you can't be sure...

Work is shape-shifting around us and this book tries to tell you how, where and why.

The latest official figures for the UK (2009) suggest that there are **3.7 million** home workers in the UK, an increase of **21% since 2001**[4]

Chapter 2

The promise of the Digital Workplace

Doctor, I think my office may be dying

If it's no longer *where* you work but *how* you work, then this shift brings control, influence, empowerment and autonomy, to each of us.

This book is about the Digital Workplace and how it exists and functions in contrast to the Physical Workplace. We all know and recognize the Physical Workplace only too well, taking for granted its longer than 100-year history of bricks and mortar (or concrete, steel, glass), housing offices, factories, warehouses and retail outlets.

Admittedly, over the decades, Physical Workplaces have become much better thought out and designed. Enter a factory today and it is a far cry from the same facility 50 years ago: better laid out, cleaner, safer, with fewer people doing routine jobs. Many offices have evolved from dull ranks of desks, populated by uniformly dressed people plonked in brutally bland rooms, into funky open spaces, plush with cafés and chill-out areas. Google has probably created the ultimate in non-traditional workplaces at its Mountain View, California centre, which boasts pool tables, climbing walls, massive aquaria and reclining couches, all designed to help create an atmosphere in which creativity can flourish.

But the Physical Workplace has peaked and is now in terminal decline. Fewer and fewer staff need to go in to the Physical Workplace any more, and the size and occupancy levels of buildings are being reduced all the time as less and less work needs to take place within the confines of the traditional Physical Workplace. In years to come, will we look back at the TV sitcom *The Office* and view it as a fascinating anthropological insight into how offices used to be? They might never totally disappear but they are already locked in an unstoppable downward spiral, becoming smaller, more flexible and fewer in number. No one likes them much and no one really needs them any more.

The Digital Workplace is the technology-enabled space where work happens – the virtual, digital equivalent of the Physical Workplace.

Work is shifting

But work is not disappearing. If anything, it is more present than ever in people's lives. We have seen a big rise in the number of women in the workplace, longer working hours and waves of whole new industries being created across the world. So where is all this work taking place if the Physical Workplace is dying?

Where it is moving to (and has already done so in varying degrees) is the Digital Workplace. This is the virtual, digital equivalent of the Physical Workplace. In fact, this is already where most work happens. The term 'Digital Workplace' was coined by Hewlett-Packard, the printer and technology manufacturer in 1998 as a move to a new style of printer. The term was left largely unused until 2010 when my colleagues and I adopted it as a useful way to describe the larger world of workplace technology. I define the Digital Workplace as the technology-enabled space where work happens. It involves all the tools we use to do our jobs: email, phone, text, intranet, micro-blogging, Internet, Office documents, shared documents, teleconferences, video, software packages, smart phones, tablets and the Cloud. These days, when you turn up to your company-owned office, factory, warehouse or shop, you will most likely be working in the Digital Workplace. But you will also be in the Digital Workplace when working from home, on the train, in a bus, taxi, restaurant, café, hotel, your sister-in-law's holiday home or on the beach – anywhere you have a phone or Internet 'connection' in fact. While the Physical Workplace is limited, restricted and bounded, the Digital Workplace is much more flexible and present – and is increasingly available and accessible anywhere, at any time.

Most people, in most organizations, including those on factory floors and in retail environments, now work at least partly in the Digital Workplace, and many people in administrative or managerial roles work pretty well exclusively in the Digital Workplace. There'll be more on when and where the impact of this is felt most strongly later. Of course, there will always be areas of work in which the reach of the Digital Workplace will continue to be limited, such as dry cleaners, hairdressers and restaurants – but even in these we can witness some digital work practices creeping in.

Free at last!

The switch to the Digital Workplace is a blessing for the world of work. The tedium and misery that still equate to work for many is facing a sea change. The reason is that the more we work in the Digital Workplace, as opposed to the Physical Workplace, the more we begin to experience unexpected benefits. These speak to our essential nature as human beings and, because of this, are highly prized.

If it's no longer *where* you work but *how* you work, then this shift brings control, influence, empowerment and autonomy, to each of us.

- You decide where you work.
- You decide, with some constraints perhaps, when you work.
- You are judged more on what you output and less on being seen and heard.
- You have to be trusted more as your boss can't 'see what you are doing.'

One sales manager I know shifted from five days in the office to working wherever he wanted. This turned out to be two days a week from home, one in a local serviced office space near his house, and two days a week at the old office. He said it 'transformed' his life as it released him from the tedious commute, allowing him to shift his hours so that he could take more exercise in the day and, crucially, he saw his children far more than before. Add those factors together and the additional value and sense of liberation is hard to quantify.

I have heard hundreds of stories like this. We are in a period where work is being made 'whole' again. I like work and enjoy the feeling of producing value. My issue was, and still is, with the nature of most work in most organizations. But we have a unique opportunity to transform – not just change or improve – but *transform* the quality of work and, in consequence, the lives of millions of people. The manager mentioned before is happier, more fulfilled, healthier – and his family is richer in non-material ways as well.

> *"Far and away the best prize that life offers is the chance to work hard at work worth doing."*
> Theodore Roosevelt

The blind leading the blind

There is a huge problem for the leaders in most organizations of all sizes. The Digital Workplace is emerging around them… 'The guys in accounts want to work from home one day a week. Can we start using micro-blogging inside the marketing department? Can we get smart phones with intranet access? Are we all on a shared marketing system? Can we just use laptops? Can we use our own iPads?' Each decision or development constitutes another step into the Digital Workplace – but virtually every company I know says that this is happening ad hoc, with no overall strategy in place.

Staff want the Digital Workplace and the technology allows it, but leaders have little idea what this really means for the shape, structure and relationships of the organizations they manage. They are wandering into the future because they are powerless to do otherwise, due to staff power, social changes, technological access, environmental impetus and legal imperatives, but they have no real clue what their company will look like in five... ten... fifteen years' time. And that is dangerous. If you don't design your future, it will design you.

So through a 'perfect storm' of technology, expectations, legislation and ecological factors, the Digital Workplace is a concept that is already here and will continue to evolve and shape work throughout the 21st century. Most people in most organizations will feel more valued and will enjoy their work more as a result.

However, as yet, the possibilities offered by the Digital Workplace are not being fully grasped by those who need to understand, enable and design this future world of work. At present, the perception for many is that the Digital Workplace simply amounts to 'a bit of home working' or a chance to reduce some office space. But those leaders and organizations that properly seize the moment will benefit, not only in efficiency and staff satisfaction, but also commercially.

Meanwhile, those who don't get on board will find themselves being swept along until one day they will stop and look around the office and wonder where everyone has gone.

If you don't design your future, it will design you.

In a nutshell

I wrote this book to say:

- Most work in most organizations is awful – and we all know it.
- The Digital Workplace offers an inevitable change that transforms work.
- Most leaders and companies are blindly wandering into the future.
- If you take the trouble to understand the Digital Workplace and design it into your future, you will be among the winners.

And there is more to come outside of the 'world of work.' The Digital Workplace will not only change the nature of work but will also change our societies – because it will change where we work, how we manage our close relationships and where we live. These are huge demographic and social issues, which will be important not only to managers but also to politicians, economists and social theorists – not to mention architects and house builders.

Chapter 3

Working the IBF way

Creating and building IBF as a technology research leader with a Digital Workplace model at its heart has been helpful for us as we have been able to pilot and pioneer new approaches that our 150-plus customers (all large enterprises) find fascinating and informative. We can try things on their behalf.

Some of the facts relating to IBF are intriguing:

- We have never had an all-hands face-to-face staff meeting since we began in 2002 – although people do meet physically.
- The two regional managing directors work together on every aspect of the business – with total trust in each other – and yet they have only met twice.
- We don't have an intranet as such but use a range of different flexible technologies and applications – all highly secure, stable and cost-effective.

'What Would Google Do?' by Jeff Jarvis

Jarvis investigates the new business model revealed by how Google operates. Instead of closing down and protecting what we create, the book explains the power and value of linking and connecting data. For me this book is about the importance of operating your business free from assumptions and habits that just 'ape' large traditional corporates. For example, there is an assumption that people must meet constantly to build trust. This is simply untrue in my experience. Meeting too frequently in the Physical Workplace actually damages trust and creates institutional behavior.

- We adopted audio and web conferencing from their earliest days – but have only dabbled with video conferencing as, in our experience, it adds little.
- We started with enterprise micro-blogging and pioneered the community aspects for Yammer (the enterprise micro-blogging tool) shortly after Yammer launched and have found it binds IBF culturally.
- On any given day, as CEO, I have no idea where any of the staff or freelance team actually are – and don't care – but I do know exactly what people are delivering, when it will arrive and what the quality of their work is. Problems surface fast and are resolved at once.
- We may 'sound' like a flat structure but in fact we have a clear hierarchy. I am the CEO and we have a four-person management team with staff reporting to them. Each person has clear reporting and budget lines. We do annual performance reviews and six-monthly check-ins, almost all conducted in the Digital Workplace and this works fine.

When will we meet again?

We have few physical meetings within the company but we do check in very regularly with each other. I expect all the team to be 'present,' which means visible in some shape or form online – Skype, Yammer, email, text, teleconference, WebEx, or so on.

We have 'IBF days' in London when anyone in the team can drop in physically to work from a central London flexible office space and anyone can work there whenever they want; people use this as it suits them. Personally, I like to work from a co-working space once a week.

The Digital Workplace is cheaper, faster and simply more fun than an office. When travel chaos hits, we are unaffected… and the Digital Workplace does away completely with that brain-numbing commute, unless of course you choose to take a trip to a café or co-working space or to meet a colleague. Everyone has more time for productive work.

A survey of 6000 UK civil servants concluded that people working **10 hours a day** were a staggering **60% more likely** to have heart disease or a heart attack than those who only worked 7 hours a day[5]

Less work and more play

Several people in the company work four days a week; the 4:3 ratio of work to play is healthier for human beings than the more usual 5:2. People on this routine can deliver far higher levels of output than bored 'five days-a-week' office people. Personally, I never work in the evening or at weekends and seldom (aside from emergencies, which are thankfully few and far between) receive or make calls outside office hours.

This causes a ripple effect in IBF and, generally, others do the same. It's entirely their call when they work, but they must never expect others to be available at odd hours just because it suits them.

Our wonderful Chief Knowledge Officer, Elizabeth Marsh, lives and therefore works from Leamington Spa. She loves playing the flute and is a member of an orchestra, so to accommodate her practice and playing, she works quite odd hours. She likes to work mornings until 2 p.m. over each of five days although she is paid for a four-day week. Frankly, I am never quite sure when and where she is working but, goodness, can she deliver and think creatively when needed, always at high quality.

Oddly, small organizations seem to be as resistant to change (and in some cases more so) than larger entities. So many small companies think they have to control everything. Recently, I was with the CEO of a small advertising agency who refuses to have anyone work from anywhere other than the office ever, basically just because this is what suits him.

No one sees how much time I'm putting into work, or when I'm working; it's just basically down to the results I produce.

Case Study

Working for IBF: Angela's Story

In 2006, IBF was interested in getting more into the North American market and Paul Miller was looking for someone to help get things started. I actually wasn't really ready to go back because my kids were very little and I wasn't sure if I would be able to make it work.

But it did seem like a very good opportunity and intriguing because it would allow me to do some career-enhancing work without having to go into an office or commit to a full-time job. It sounded like something I could ramp up or cut back on, as and when I needed to; it sounded ultimately flexible. So I had a quick call with Paul who sold me on the whole idea and I took it from there.

Previously, when I worked in a corporate office, I was usually in a cubicle somewhere. I did like being able to interact physically with my co-workers but found in a lot of cases it was more or less the same as virtual working. I would have teleconferences with people in the building next door – or sometimes it would just be a couple of us in cubicles on the same teleconference but not the same room.

Room with a view

Now I'm in my home environment where I can get up and do something I need to around the house… and I have a beautiful view outside my window, instead of being stuck in a windowless cubicle.

One thing I really like about working for IBF is that we're not together physically and no one can see how much time I'm putting into work, or when I'm working; it's just basically down to the results I produce. We don't care where people are working from, or what time of night or time zone they're working in, as long as they're getting the job done. You're just accountable to yourself and the project you're on.

We have team members working from all over the place. One editor is based in Grenada, and one of the evaluators lives in northern France – IBF virtual work supports her passion for mountaineering. One guy spent a month in South Africa following the World Cup and continued working as usual. If he hadn't told me where he was working from I would never have known.

We also have team members who travel around the world, while working virtually and supporting their travels with IBF work. A lot of phone calls and teleconferences start with the question, 'So, where are you working today?' It's interesting to hear and inspires me to get out and about as well.

Divided loyalties

But working at home can be challenging and there are some ground rules I think each individual has to set for him or herself. I always have my iPhone on me and I check it periodically. But if I'm just sitting with my kids, and not actively doing something with them, it's all too easy to grab the phone and get sucked into 'continuous partial attention.' Your mind is kind of split – physically there with the family but with one eye on your emails. There's a Skype popping in the background… you send a quick text to somebody… you're sort of there, but not fully so.

Getting to know you

There are challenges also in getting to know your co-workers in the virtual world. You have to make a bigger effort to get to know someone when you only interact via phone or Skype. For example, I don't always pick up on some people's humour.

Perhaps we'll be on a conference call and others will be laughing about something and I just don't get it, especially very dry humour. If I haven't spent much time with these colleagues, either one-to-one or on virtual calls, and I don't really know their personality, it can be hard to pick up on subtleties.

Although I have worked virtually for IBF right from my first day, I had always felt a little on the outside – that is, until they announced that the physical office space was going completely. Now we're *all* virtual. Even though very few people actually ever went into the IBF office, I did feel a little remote, although when Yammer was introduced it helped a lot in terms of instilling a sense of inclusion. Nevertheless, there was a mental switch for me when the whole team went virtual and I did feel 'Okay, now we are all on a level playing field. Even if somebody is physically in London or New York, they're just doing the same as I am from Ohio. It doesn't matter where they are!'

Chapter 4

The right people – in the right situation

At IBF, whether it be as staff or freelance, we only ever bring in fantastic people. If you are not deeply intelligent in a 'business useful way' then you are very unlikely to get hired and you are certain not to last. If you only have the very best people, the effect is endemic; they create this quality in others. So many organizations of all sizes have people who are mediocre and should not be employed there. You can train and develop to your heart's content but, basically, poor staff will remain poor forever in my experience. The rest is just window dressing.

"Employ staff who can do the job better than you can do it yourself. Then be grateful they can do it better than you."

Simon Carter, fashion retailer

Less political, more effective

We stamp on any politics within IBF quickly and directly. I have taken direct action with everyone I can think of in the company at some time – and yet many people have been with the business for years and we have little turnover of people; it works. I have also been told bluntly that I have a tendency to generate too many ideas, which can confuse the business, plus I can be a bit moody at times. Well, I'm afraid that's me, but I do try to keep both tendencies in check.

An interesting observation is that the only really tricky personnel issues we have ever had – personality clashes and the like – have happened in the one physical office we used to have. Offices cause friction because people don't actually want to be there every day. The Digital Workplace is in fact easier to manage and I'm convinced fewer 'personnel issues' arise. I have no statistical evidence for this but it is certainly what I have observed.

And how do we find the right people? We have customers who have become freelancers and then joined the staff, as well as staff and freelancers who have gone the other way and become customers. There is a fluid movement of people wearing different hats around IBF.

This creates a healthy and commercially strong dynamic in what we call 'IBF World,' a community that includes customers, suppliers and the team, as well as prospective customers with whom we are just building a relationship.

'Good to Great'
by Jim Collins

This book conveys key precepts of successful business. The part that has always stayed with me was the vital role of getting the best people onboard and dispensing fast with anyone who cannot operate at a high level. IBF has amazingly high quality and high intelligence staff and freelancers, and anyone less than that finds IBF life unbearable. This is vital. Many organizations waste years trying to encourage, train and performance-manage low-grade staff to become at least average employees – and mostly fail. Don't waste time with them; let someone else have that burden.

IBF standards and mantras – otherwise known as 'how we work'

- We never write down our business purpose or vision. People should be able to 'feel' what the business is about and practise that day in day out. If you need to write it down, then you have no clear purpose or vision.
- We don't swear in IBF. Not because I disapprove of swearing but because it demeans the business conversation. It does happen, but hardly ever. I have at certain times worked with people who swear a great deal and observed that it damages relationships within the business; in the end no one knows what is acceptable or unacceptable in language.
- We apply similar standards in all our dealings with our members (or customers) and they enjoy being with each other and the IBF team. It's an ethic and practice that everyone can feel.
- Ironically, for a company called the Intranet Benchmarking Forum, we don't have an intranet… but we are getting to the stage where we could really benefit from one. This is a weak point for us but we have used around 20 different workplace technologies and apps on various desktop, mobile, laptop and iPad-style devices.

- We have very well documented on-boarding processes and everyone (both employees and members) know exactly what is happening from Day One of their association with the company; they get to know the team fast through collaborative work.
- We take every opportunity to meet physically but tend to attach these meetings to customer gatherings or sales events, that is, to coincide with pre-set services we are delivering.
- I have weekly or monthly calls with every key person in the business and feel connected and close to the many IBF people, some of whom I have never physically met.

Calculating the true number of homeworkers in the US is very difficult, but Telework Research Nework estimates:

- **20–30 million** work from home at least one day a week
- **15–20 million** are mobile workers
- **10–15 million** home businesses
- **15–20 million** people work at home part time
- **3 million** work at home full time[6]

Top ten maxims for the Digital Workplace

1. The Digital Workplace enables people to have power and influence over *how*, *where* and *when* they work.

2. Work is shifting rapidly from the physical to the digital – and the progress is steady, fast and relentless.

3. The office as we know it is in decline – but what is coming in its place? As yet unclear.

4. The Digital Workplace has the ability to transform positively the nature, design and experience of work for most people in most organizations.

5. My own organization experience holds lessons for large enterprises embarking on moving deeper into the Digital Workplace.

6. You do not need to meet physically to build trust; while it can help it is not essential.

7. If you don't design your future in this area, it will design you.

8. The Digital Workplace produces less 'politics' and distractions than being in an office all day, every day.

9. Work itself is not getting easier or less demanding – it's just that its 'shape' and location are changing.

10. It is impossible to implement a successful Digital Workplace without high degrees of trust and autonomy.

Close

WORK WORLD

Chapter 5

Leadership

Leaders have nowhere to hide – thank goodness!

When I was a speechwriter for senior executives back in the late 80s, in the pre-Internet days (hard to believe I know), the then Chief Executive for Citibank in the UK, confessed to me one day that it was virtually impossible to lead the organization. It was mayhem within the bank and individual departments just 'did their own thing.'

Leading modern organizations was then – and still is – a crazy process in which companies like to exude an air of confident stability, masking the chaos, inefficiencies and duplication that are daily realities: internal strife, massively variable staff capabilities (organizations are filled with mediocre people who it is virtually impossible to remove), power politics, lack of true innovation… and on it goes.

'Henry V'
by William Shakespeare
(my interpretation)

When King Henry needs his troops to be at their best, he speaks passionately and directly, and in his own voice to them. He does not ask his advisers to massage his message. He does not avoid tough questions because his number two said it was too risky. He does not delegate the task to his version of Corporate Communications. He takes his role as the only true communicator seriously and uses his own authentic voice to bring out the best in his troops.

Anyhow, thankfully, this book is not tasked with creating a panacea for all organizational ills. Instead, in this section, it asks the question 'What does the Digital Workplace offer for the modern CEO and their C-level team?'

Use everything you can

The Digital Workplace offers a host of goodies that should have leaders falling to their knees and thanking the heavens they are in the top slots today rather than during the 80s.

Take the story of former BT (as was British Telecom) CEO, Ben Verwaayen. According to the serial blogger and early web proponent within BT, Richard Dennison, the way Ben used online question and answer sessions to communicate with staff transformed the culture in the then rather staid BT, turning it into a highly engaged place to work.

Here are Richard's tips for leaders based on what he saw Ben do:

- Sit at the computer yourself and type the answers to the questions in your own words... if they don't come across as being authentic, you'll look like an idiot who doesn't get it.
- Don't surround yourself with internal communication managers and PR people advising on you on what to say – say what you believe and say it with passion.
- Don't let some PR or communications person screen out questions before you see them –

you should see all the questions, including the abusive ones, even if you don't choose to answer them. If you don't know how people are feeling you cannot have an authentic conversation with them.

→ Keep your answers short and simple and don't write meaningless tripe!

What Ben did was spend an hour a month (that's all it took) and engaged in real live chat conversations with people from anywhere in the company via the BT intranet. How easy is that? What an efficient use of his time. What a powerful a way to connect to the CEO.

'Not my style' – that's no excuse!

I was hosting an episode of Intranets Live, IBF's monthly live broadcast of live intranets and Digital Workplaces in 2010, and I recall an exchange with a supposedly senior communications agency expert from the US who, when faced with the question of senior leaders and use of social media, 2.0 and all things interactive, said: 'It suits some execs but not others and that's fine.'

I countered this by suggesting that any C-level executive, and particularly any CEO, who does not lead through technology, ought not to be leading a modern advanced organization. Using collaboration tools, live chat, video and audio messages, recordings and so on, is not just an option, it's an essential.

Any C-level executive, and particularly any CEO, who does not lead through technology, ought not to be leading a modern advanced organization.

When just elected, President Obama appointed the country's first Chief Technology Officer, closely followed by a Chief Information Officer. In so doing he announced that he understood that technical leadership was all part of being President.

But what if you are quiet and introverted? Either that's tough and you are perhaps unsuited to leadership (and I am struggling to think of successful CEOs who fit that description entirely), or you need to push through that limitation and discover an avenue that works for you.

Maybe you can jump into discussions happening online in the organization and add your own comments on changes to the pension system or a brand extension that has got people in a stew; or you might enjoy talking live with a face-to-face audience which can be recorded and posted online to allow people to post their comments and questions.

The best example of a successful CEO who is not exactly a charismatic extrovert has to be Bill Gates. I have heard him talk and he is a quiet, thoughtful guy, but he makes his points strongly and he does not hide away. At Microsoft he would from time to time send all staff emails and people would, of course, read them carefully; and yes, it was a few years ago, but he was a leader in the Digital Workplace.

A blessing for any modern leader

In fact there is no option. If you want to play the invisible leader, then staff will just not 'get you'. I have heard so many people moan when they have a CEO who is absent from view. People talk, they begin to complain and, most importantly, the discretionary energy that the staff bring to work each day (the drive and ambition that they can choose or very often choose *not* to include in that day's work) is left outside. You still pay the same money but you get less of everything. Leadership can inspire people, and Ben at BT understood this completely. Ditto, John Chambers, CEO at Cisco; and Peter Appel, CEO at Deutsche Post/DHL.

It's smaller scale, but as CEO of IBF, I love using the Digital Workplace to lead. Personally, I dislike physical travel (airports, taxis, planes, bags, hotels and all the rest) but I love 'digital travel.'

Globally it is estimated there are a staggering

5.3 billion

mobile phone subscribers, equating to 77% of the population. In 2011, over 85% of new devices were web-enabled. In the US, 25% of people who surf the web do so only on their mobile. It has been estimated that by 2014 there will be more people accessing the web from their mobiles than their desktops[7]

So, each day I will post a comment or 20 on various Yammer micro-blogging channels to let people know what I am up to – perhaps settled working in a café near Camden, or in the shared office space we use on certain days, or elsewhere – plus pass opinions and forward prompts to the

Leading in the Digital Workplace is fun, fast, cheap, flexible, effective and keeps your energy fresh.

staff and freelancers globally.

We hold bi-monthly live teleconferences during which staff update each other on what is going on and we discuss where we are heading (all recorded for those who can't make the live calls). There is a marked lack of physical meetings; instead, we keep up to date through a steady stream of check-ins via Skype, WebEx and Live Meeting.

One day recently, I emerged from a meeting with an IBF member in Chicago and spent a lovely two hours in a restaurant garden on the superb iPhone 4. I posted to Twitter, Linkedin and four Yammer networks; sent several emails; had three Skype calls and made a couple of mobile calls. Some of the communication was 'within IBF' and others with customers and prospective customers – often the divide is quite hard to spot. Leading in the Digital Workplace is fun, fast, cheap, flexible, effective and keeps your energy fresh.

More or less power for the CEO?

The Digital Workplace has not changed the power structure in the company; as CEO I have no more or less power now. What has changed is that the Digital Workplace has exposed the true character of the company and made its power flows more explicit. If there is a discussion happening in the Digital Workplace, then your participation or absence is always visible. You can choose to stay out of the conversation, as I do when the benchmarking team are chewing over the finer points of usability, or jump in at once when a member asks whether the arrival of Microsoft as a member means IBF is compromising its independence. But whatever you do or don't do is out there. You have nowhere to hide as a CEO – and nor should you. That's your job as a leader.

So, you can gain in real visibility as Ben Verwaayen showed at BT through live online chats. Other good examples are active blogger Tom Glocer, until recently CEO of Thomson Reuters, and micro-blogger Marc Benioff, CEO of Salesforce.com. All three have embraced technology and have not shied away from entering as peers a dialogue with those whom they lead. It might 'feel' like Marc, Tom and Ben are 'just one of the guys' but the truth is that no one ever forgets they run the show and, in Marc's case at least, also owns the show! If anything their power (rather than control) is extended by people being able to witness for themselves how they think on their feet. Control is harder to retain, but leading through power, influence, impact and reputation is far more effective than trying to control.

Even giants can change

When an organization shifts from prioritizing control to concentrating on results (in other words, from policing to enabling, from Physical to Digital), this can have a very liberating effect on the culture of the company. Towards the end of the 20th century, IBM was a strong, but arrogant, giant. Then it hit a wall and performance plummeted. In recent years, it has opened up to dialogue and is listening at all levels, and today it is once again respected as a huge – but now healthy and well-functioning – giant. The command and control ethos of the past has been replaced by an open, collaborative culture and this has revitalized IBM as a rewarding place in which to work.

"Just remember, an informed CEO is a happy CEO… I want an instant snapshot of what's going on in the minds of my people."

Michael Critelli, former CEO, Pitney Bowes

Case Study

Michael Critelli's challenge at Pitney Bowes

How do you lead your company through extreme change while simultaneously establishing a solid 'one firm' culture?

That was the challenge facing Michael Critelli. As CEO of Pitney Bowes from 1996 to 2007, Critelli faced an uphill struggle going into the new millennium. The firm traditionally manufactured equipment for physical mailing, but the predictions were that digital technology would kill the industry.

The firm was also highly acquisitive; from 2001 to 2006 it swallowed up more than 50 entities, a nightmare to assimilate both logistically and culturally.

How did Critelli utilize the Digital Workplace to lead and strengthen Pitney Bowes towards becoming the US$5.6 billion, 33,000-employee global business it is today?

The fact that Critelli majored in communications and appreciated the value of dialogue with small, defined interest groups meant he had the right mindset for developing a sophisticated internal communications strategy, even at a point when the technology was still embryonic. He also actively avoided being an 'anonymous' CEO, seizing every opportunity to make sure his message got across to everybody at Pitney Bowes and that they were aware that this came directly from him.[8]

Messages and metric

A good example of the kind of initiative Critelli used was one introduced after 9/11. Four employees of the company had been killed in the attacks and Critelli felt the company needed leadership. Almost immediately, he instigated weekly audio messages from himself (called 'Power Talks'), which

all employees could listen to on voicemail or catch later via the intranet.[9]

Critelli and his team then used the statistics on who was listening to or deleting these voicemails to make changes for more effective distribution. For example, they found that the company's call centre workers tended to delete the messages on a Monday morning. Digging deeper, it was discovered that these workers often had a backlog of voice messages from the weekend and yet another voicemail was the last thing they wanted to receive. By sending the Power Talk on a Wednesday instead, the number of deleted messages was dramatically reduced.

Once the Power Talks were established and had been optimized, they continued. Critelli went about applying the same mix of targeted messages and metrics to other communication methods as well. Roadshows aimed at unifying company culture were streamed live on the intranet; viral videos were used to promote internal healthcare initiatives; and detailed analysis of electronic employee engagement surveys was made available.[8]

The Digital Workplace in Pitney Bowes

Although he retired from Pitney Bowes in 2008, Critelli continues to use the Digital Workplace through his blog, 'Open Mike' (www.mikecritelli.com). He started this while at Pitney Bowes but now uses it to report on his own activities as well as issues such as healthcare.[10]

Meanwhile, the foundations Critelli laid for a culture of using the Digital Workplace at Pitney Bowes have continued under the new CEO, Murray D Martin. Further initiatives have included: PB Voice, an area on the intranet that allows open questions and answers with senior management; a Yammer community with the emphasis on learning and development; and an online innovations forum called Ideanet, which has so far resulted in 500 ideas being adopted into company processes and products.[11]

Chapter 6

Teams

Oh, how we long for the good old days! Remember when a manager could scan the office and see at a glance the good folks who worked for him? Maybe it's fiction, but Lord Sugar, star of Britain's TV series *The Apprentice* and hyper-successful businessman, is reputed to have arranged his early offices so that he was seated in the middle on a podium from whence he could see everyone and bark instructions at them. Managing teams used to, at least superficially, feel so simple. You worked physically with those who reported to you and could easily monitor what they were up to, the hours they worked, how many calls they were making, and so on.

> *"In the future, physical meetings will take on sacred, almost mythic proportions."*
>
> Timothy Leary – LSD guru and Harvard Professor (1990)

These patterns can be extremely hard to break. I had an alarming conversation recently with one major financial services company that was embarking on a brave new mission to create a 'new digital working environment' for the company, led by the CEO. There would be comfy chairs, the latest gadgets and an 'agency feel' to an entire floor in their HQ. 'And you'll be letting people work where there they want some of the time – from home, in cafés, on the move – you'll cut down on the commute?' I asked. 'Oh no, everyone will still have to come in every day.' What's the point of that? You treat them like employees from the 1960s but then expect that just because they're sitting on beanbags they will somehow morph into 21st century digital natives. What nonsense!

Managing virtually

In my experience, managing people virtually is far easier than managing people in person. Now, this could be a personal limitation in me... well, it could be, except that it appears to be true for other managers in IBF too. Are we just unique or odd? Perhaps we just happen to be part of an industry that lends itself to this way of working and is therefore less prone to issues caused by having no office? It's hard to say, but I suspect this is not the case. It is not so much that managing is easier when you see people less often but rather that forcing people to be together physically day in day out tends to cause friction. Traditional workspaces seem, by their nature, to run the risk of cultivating disagreement, tension and personality clashes.

If we meet only when it is useful for us, then we have power and influence over that meeting. Times are agreed mutually as part of work and, instead of being 'stuck' together every day, we can meet when we choose and need to. This engenders more respect and creates fewer opportunities for problems to arise. Colleagues in any company will form cliques and friendships – some people you connect with and others you don't – that's human nature. But offices also force proximity and irritation, disharmony and power games.

If we meet only when it is useful for us, then we have power and influence over that meeting.

Middle managers are running scared

Two groups appear to be most threatened by the Digital Workplace: managers who are not used to managing remotely and have had little or no training in how to do it; and workers who are so accustomed to their desk and office that the very idea of working flexibly two days a week feels as unsettling as losing their home. To counter this, Dutch finance group SNS REAAL has addressed the issue by paying attention to training managers in how to manage people they see only infrequently.

Managing in the Digital Workplace requires a different approach. Employees still need to be managed but this can't be done by policing. Results are all-important but it is not necessary to watch the production process.

If I see you just one or two days a week, or less, then what we set out to achieve on the days we are together must change; and when we are physically apart we nevertheless need to stay connected and available. In IBF, we expect presence online through technology but not physical presence. Employees can work wherever they like but we need to be able to interact with them during normal working hours.

In some ways the hardest thing is to shift those workers who have become addicted to their desks. Old Testament stories show that when Moses led the Jewish people out of Egypt to freedom from slavery and towards the

Managing in the Digital Workplace requires a different approach. Employees still need to be managed but this can't be done by policing.

'Promised Land' it was not all smooth sailing. Out in the desert, the Jews grew restless. Yes, they were glad to be free from slavery, but at the same time they moaned that being free was all well and good, and yes, they did want to reach the Promised Land, but at the same time they missed the shelter, the food and the security they had been used to in Egypt. In non-religious paraphrasing, God through Moses said: 'Fine. You can wander in the desert for 40 years. Clearly, we can take the Jews out of Egypt but we can't take Egypt out of the Jews.' A generational change was needed and Moses waited until the children born in the desert were old enough to enter the Promised Land. This is a story of change – one of the earliest and one of the best. It takes time and some groups will never be able to make the journey; in which case, give them a taste and if it doesn't work, then wait until the workforce demographics change.

IBM: Blue Pages

This really led the field in its day a few years ago. The killer intranet app was always the employee directory. Organizations never had up-to-date directories that told you everything you needed to know. The IBM 'Blue Pages' did just that and actually connected people with people in a way everyone had dreamed of. It was a sort of 'Facebook for the enterprise' in the days before Mark Zuckerberg had even started at Harvard!

How to avoid some being more equal than others

One responsibility of the virtual manager that can often be challenging is successfully mixing the physical and the virtual. Meetings in which some people are there in person and others present virtually tend not to work well.

The digital attendees almost invariably feel marginalized, irrespective of any efforts the 'in person' group may make to include those not in the room. We have tried various different ways of dealing with this and have found that the best approach is to ask those who are physically together to dial into the teleconference from separate quiet areas of the building. In this way a level playing field is created for everyone attending the meeting.

Trust – employers believe their staff are cheats!

Trust is an absolutely key issue. The financial company I talked about earlier displayed zero trust in their staff. So, while they were playing lip service to the Digital Workplace, they still wanted to watch everyone, all the time. In a properly functioning Digital Workplace, people who work together find they need less and less to be physically in the same room. Your boss and your colleagues can be anywhere and most likely are.

In IBF, 90% of the staff never came into our one physical office, so we got rid of it. Our employees work on several continents, mostly from home. We meet physically at customer (or member, as we call them) meetings and use these times as opportunities to say hello, rather than to work together. We can already work effortlessly in a virtual digital world. Three years had passed before our global and US managing directors met in person. They had already navigated every

"Accept the fact that we have to treat almost anybody as a volunteer."

Peter Drucker – business theorist, highly influential in shaping modern management work experience possible together: hiring, firing, performance reviews, customer pitches, work delivery, financial management and so on. They already felt they knew each other very well before they even met; the physical meeting when it eventually happened, while pleasant, actually added very little to the rounded relationship they had already developed.

> To my mind, the most delicate part of digital working is managing people. It takes a conscious effort to build personal connections.

Managing in the Digital Workplace requires persistent effort and attention. It is impossible to manage virtual teams without higher levels of trust than have been the norm.

When you cannot see what people are actually doing, you have to trust that they will work well and, crucially, you must judge them based on what they *output* rather than what they *input*.

We have an Online Services Manager whom I have never met and probably never will – unless she comes for Christmas drinks – but I notice that items are promptly loaded to various sites we run and I hear positive comments about her from colleagues. She is carefully monitored for performance and delivery but, now that confidence in her is established, her manager has stepped back and just checks in with her via Skype once a week. On the other hand, this virtual staffer is digitally very present in our Yammer micro-blog staff network; she also posts to our blogs and regularly enters into various online meetings. If she were physically in the office, what would it add in tangible business terms?

Case Study

Working for IBF: Nancy's Story

I've been working with IBF for nearly five years. I'd say that 90% of my work is delivered remotely. I just love it. I have the right balance of digital working and in-person interaction, whether this is by leveraging video chat for a team call or hopping on a plane to San Francisco for a few days of concentrated time with the team and our members.

During my Wall Street days, I had a two-hour commute each way. Commuting was effectively an extension of the already long working day and I used the train journeys productively (catching up on emails, calls, research and so on). I shudder to think how many PowerPoint decks I churned on those trains!

Today, my 'commute' is more about making a mental shift rather than a location change. At the start my working day, I light a candle and close my eyes for a moment to visualize what needs to get accomplished in the day. This one-minute meditation effectively enables a mindset shift from home life to work life and an environmental shift from a family space to a workspace.

I developed this meditation in the late 1990s, when I became the first virtual employee at JP Morgan Chase. As an early adopter of digital working, I had to figure out how to create a physical workspace that was conducive to working as opposed to the temptation to do the laundry. Bear in mind this was a point in time well before most companies had work/life balance programs.

To my mind, the most delicate part of digital working is managing people. It takes a conscious effort to build personal connections, so that the trust base is strong when issues arise. Not everyone readily verbalizes stressors or challenges. Traditional non-verbal cues need to be surfaced quickly and proactively.

Within the IBF team, we have a variety of check-ins with each other. A few examples include:

Virtual water cooler: We use Yammer as a virtual water cooler, so people can share what they are up to day-to-day. This is a great way to connect individual efforts with each other, our members and even prospective members, as well as to share new ideas, challenges, and even plans for the upcoming weekend.

One-on-ones: We also have regular one-on-one check-ins to connect on a more personal level (e.g. to draw out how someone is doing, to understand and work through challenges that have come up, to collaborate on new ideas, to work on professional development plans, to share timely performance feedback, to strategize on a problem).

Real-time exchanges: We rely heavily on instant messaging for real-time exchanges in the course of the working day. It's a great way to address a quick question, enlist a colleague's help if you are stuck on something or to give an update on a time-critical deliverable.

Collaboration spaces: We also leverage various tools for real-time collaboration on projects and service delivery.

A digital-minded culture, coupled with an entrepreneurial and adaptive organization, gives IBF distinct advantages, including:

Motivated staff: The level of motivation and productivity of the staff is high. I think this is in part because we use freelancers who are very motivated to contribute to the team. The notion of complacency never seems to set in

because we choose freelancers who are accustomed to working digitally and are results-oriented.

Higher levels of trust: Trust levels run very high inside IBF and that's no small accomplishment. Office politics simply aren't a factor; neither is face-time. The culture is one that supports direct, honest feedback on a timely basis. One never has to guess how or where things are going. These are essential elements to cultivating a trust base.

Experimental ground: Because of the nature of our business we tend to be early adopters of new technologies and approaches to work. Treating IBF as experimental ground allows us to bring good practices to our members.

Quality of delivery: IBF and now the Digital Workplace Forum (DWF) are entrepreneurial and highly adaptive businesses. When I stop to think about the throughput and quality of work our team is producing collectively, it's quite remarkable (benchmarking, IBF Live, IBF 24, IBF Knowledge Base, member meetings, research – and now DWF). I have regular catch-up calls with members and they speak volumes about the quality and impact that the IBF team is having on the performance of their intranets.

Innovation: IBF is a creative, dynamic and results-driven organization that supports experiential learning and embraces creative tension. IBF leadership is at the heart of this way of working. We have an inspired and inspiring leader in Paul Miller whose DNA imprint permeates every layer of the organization.

I think it would be quite difficult for me to go back to a very traditional corporate structure that's rooted in face-time. IBF has all the 'business as usual' infrastructure, with very solid processes, systems and people, but we also operate in a fresh, responsive and energetic digital environment. It's a unique and refreshing experience every day.

Chapter 7

Engagement

No need for badges and coffee

Famously in corporate folklore, the internal communications sage that is Mike Wing, father of IBM's corporate intranet, invented the 'IBM Jam' in 2001. This, as far as I know, was the first time any large organization had set about engaging large numbers of staff in a shared, real-time, virtual conversation. For three days, IBM staff could post their views about issues around corporate social responsibility in the company. It came after a period of pain in IBM, the turbulence of the 1990s, and was a huge success. It changed the way companies look at non-physical engagement.

Oddly, the take-up of such innovative approaches by others is always very slow. They hesitate to do similar things internally out of fear, worried that they will fall flat on their faces because their company 'isn't IBM.' That said, Aviva, the major European insurance giant, has now run several 'Aviva Days' lasting 24 hours, once a year, with impressive results in terms of staff participation. In IBF, each year we stage IBF 24, a rather lunatic marathon journey across the globe during which we host live tours of the world's best intranets and Digital Workplaces, with thousands tuning in from over 50 countries. There are things you can do in a digital sphere that are simply impossible physically.

When organizations fragment

When the organization fragments physically, it is essential for leaders to engage staff and contractors in new ways. Why? Well, there is no legal compulsion, but research by leading HR consultancy Watson Wyatt (and now supported by many others) has shown that staff satisfaction leads to customer satisfaction and enhanced shareholder value.[12] It reduces staff turnover, motivates performance, and in turn drives efficiency. Now, you can choose to ignore staff engagement... but at your peril!

So, what does all this mean in the Digital Workplace? Some companies have come up with initiatives that allow their staff to share useful experience with each other in creative and engaging ways. For instance, AXA, the French healthcare and insurance company, invites their staff to post videos of how to handle customers in call centres, while British Airways has a facility by which staff can post video clips of how to deal with certain recurring problems that colleagues may also be faced with.

AXA: Customer Service

How do you deal with a really angry customer in a call centre? AXA has encouraged staff to make and post short direct-to-camera 'training' films. The contributors choose their own words to explain from experience how to calm people down and to turn a tricky situation into a reasonable interaction. Simple, free and brilliant. Other staff can then vote on the clips and post comments to say thanks and add any other ideas.

How many people feel engaged by watching a single 'tip to help you' video? Stats would say that each video affects about 1000 people. So, post 100 such clips a year and that's 100,000 staff engaged – at zero cost. Compare that with the old-style 'Let's have a corporate conference for all the middle managers costing US$500,000!'

Nirvana for senior managers

British Airways: Coffee Clips

The BA intranet is fully mobile and has, for years, allowed staff in the flight crew to access rotas, HR, pay, pension and schedules from anywhere via Internet access. Another lovely feature of their intranet is the collection of video clips posted by staff of often mundane and yet incredibly useful stuff that other staff members might need to know on a daily basis, such as a one-minute video demonstrating how to get a coffee jug free when stuck in the small grill on a Boeing 757.

Most CEOs in advanced organizations will blog, handle live chats, post to micro-blogging areas, run videos with commenting; these simple ways of exploiting technology to touch staff work extremely well and are universally popular. CEOs can often be nervous about introducing this kind of communication in case someone asks a question they find difficult to answer. Others will give it a go but then quickly ditch the whole thing after someone raises an awkward issue. Yes, some people will post tricky questions but the issue is already out there, so why not take the opportunity to have it out in the open and address the matter? Or, let someone else dip in and answer the question? In fact, what often happens in these situations is that the questioner or commenter will be told to stop being so silly – the self-correcting mechanism of peer pressure takes effect. In IBF, I watch what is being spoken about in the digital sphere and then either praise people publicly or course-correct as needed. It keeps the company healthy and focused.

When we don't see our colleagues every day, as leaders we do need to work harder to create engagement. However, being together physically often creates an illusion of engagement rather than a reality. Just because you can see your colleagues, are you actually engaged with them? Since we have no offices at IBF, we make a point of arranging to meet in person; this requires planning and structure. However, it's the start to the day when the various online channels open up that provides our real route to connection, supplemented by regular calls and teleconferences to keep everyone on the same page.

There is an extremely strong link between the ability for employees to choose their physical location and their level of engagement.

Case Study

Wayne Clarke on what makes for the 'Best Companies'

Wayne Clarke, International Partner of the Best Companies Partnership,[13] is an international expert on employee engagement and consistently ranked as one of the most influential professionals in HR. Best Companies conducts employee engagement surveys for many leading companies and compiles the *Sunday Times* 'Best Companies to Work For' survey.

Best Companies' research suggests an extremely strong link between the ability for employees to choose their physical location and their level of engagement: *We did some specific work looking at the effect of flexible benefits on employee engagement. We looked at everything from health club memberships to profit-related pay to crèche schemes for kids. The only benefit we found that seemed to affect engagement scores was flexible working.*

Best Companies looked at two sorts of flexible working.

The first was having a choice of location – the option to work from wherever one chooses – which can bring with it communication challenges from a company perspective. The other looked at flexibility in terms of the type of work performed. In companies where a worker might do 10–12-hour shifts performing a pretty rote task, a choice of location might not be an option. But it might be possible to vary the type of work performed, and this can impact on engagement score.

The impact of flexible working as an option was so big that where companies offered flexible working options for their people, it improved engagement scores

dramatically. If you didn't do it, you'd be on average in the bottom 50 of the 700 companies we examined. If you had flexible working options, it had the power to move you to near the top 10% in engagement scores. It was that phenomenal.

If you have a flexible working culture, it means you've probably got a culture of trust... a culture where your managers get on and believe that people can work under their own steam. That says a lot about an engaged organization. In companies where there are micro-managers, who don't have confidence in their people to let them work from a location of choice, you have a disengaged workforce.

In the survey, Clarke asked 'What could make it a better workplace?' and this generated a huge number of comments that flexible working options would really help.

One of the biggest gripes identified arises when people are aware that other members of staff in the same organization have the ability to work flexibly while they don't. They hate the inequality.

Flexibility isn't just a quick fix

On implementing flexibility, Clarke says: *In departments where you've got the best people managers and the staff feel engaged, you know it's probably going to work out for them because they're starting out with a pretty engaged way of working and they already get on. But if you have a department where people fundamentally don't like the manager and the manager fundamentally doesn't like the people they work with, and you then implement some sort of flexible working, it's probably going to be a disaster because you're just going to create an even broader gap between those individuals. We say you have to fix that managerial and personnel issue first.*

Chapter 8

Buildings

One of the intriguing questions is what the Digital Workplace means in the short and longer terms for the buildings owned and leased by organizations of all sizes. This issue is now occupying the minds of senior heads of real estate across the world's major enterprises – and the questions are flowing:

- How many buildings will we need in three, five, ten years' time?
- How many people will work there and how often?
- What do they need when they are in the building?
- What is the overall strategy in this area?
- Is digital investment always better than physical?
- For every office desk we do not need, what is the accompanying digital spend required?

The trend is to reduce building stock and to assume that more and more people will work less and less in those buildings. The benefits can be seen from Microsoft's experience in Holland, Sun's pioneering 'Open Work' Program and BT's workshifting in the past decade. These have resulted in financial savings across buildings, fuel and technology; high carbon savings; and increased productivity due to reduced travel time. These shifts tick many business drivers simultaneously.

The tricky part, as any global head of real estate will confirm, is in planning for something that requires such a long lead time as building stock, in a world shaped by technology that changes by the day, alongside work patterns driven by future staff who are still at school now but who will, no doubt, make demands in five years' time that are very hard to predict.

In a report sponsored by Citrix, the consultancy Future of Work Unlimited estimated that on a global basis, approximately

26% of the total global workforce (of 3 billion)

is currently working away from the primary office at least two days a week[14]

Gone away

At IBF, our one fairly small physical office went early in 2011, meaning that the entire company of around 50 people became completely virtual. Bearing in mind the nature of the company, we have a knowledge-based and perhaps exceptionally technology-literate workforce but, nevertheless, the implications of this shift are interesting. The cost savings from having no buildings in Year One are in fact negligible, due to the extra initial investment in our Digital Workplace necessary to ensure seamless access from anywhere for all. But in future years there will be a steady saving from having no rent and related servicing costs. People still need to meet and to work, so we now rent a conference-style space one or two days a month and we also regularly use a co-working space that suits some people.

The bottom line for our fast-growing, medium-sized company is that we will never need an office again and the related needs for communication, culture, management, performance and innovation are now being covered digitally. For major organizations the pattern I expect to see emerging is a move to smaller and more flexible buildings that can accommodate far more people than could ever be in the same office on a given day. When people are in the office, they will not have their own desk but instead they will have access to connected space and meeting spaces where they can interact physically with colleagues they see only occasionally.

The average utilization of any office at any given time is estimated to be **about 45%**[15]

At the same time, corporations are likely to retain offices or 'pods' located in all their key regions, but these may be shared with other organizations of a similar size and organized as flexible co-working spaces with dedicated areas restricted to company X or Y for security.

The next tier will be even more flexed co-working spaces where the barriers between companies will be invisible; below that there will emerge a new model of hybrid working spaces, situated near to where staff live. The reality is we all need to be somewhere physically and not everyone wants to work at home or in a Barnes & Noble coffee shop every day.

Who looks after the physical environment of the Digital Workplace?

A fascinating question is whether those responsible for real estate will remain accountable purely for the physical stock the organization owns or leases, or whether their remit will be extended into other physical spaces that are only partly occupied by the company. Will they also be responsible for the home environments of their staff, for instance? And who will look after the digital worlds these staff inhabit and the services they require in order to function in this new, flexible world of work?

Helping to redesign what the devolved workspace of the digital world looks like will be the greatest challenge for the commercial real estate community in a very long while. This transformation offers huge potential financial and other gains but, right now, we are only just in the very early stages of understanding all the implications of this new way of working, not the least of which is the provision of many diverse, smaller and more flexible workspaces.

A superb paper from leading office design consultancy ARUP called 'Living Workplace,'[16] published in 2011, looks at the future of the workplace in 2030 and in the lead-up to that time. What it envisages is a new style of office, designed for collaboration, social connections and shared spaces – a world where the office has to 'fight' to justify attracting staff to be there; balancing and switching between the digital and physical worlds of work. There will be offices that include community spaces like libraries, galleries and shops; and offices where companies co-occupy with other companies to optimize the use of space.

"Open Work is really about choices for individuals. What a terrific opportunity to bring in intellectual capital not able to physically join a workforce for whatever reasons."

Ann Bamesberger,
former VP,
Open Work Services,
Sun Microsystems

The other trend taking shape is the evolution of the 'third place' that is neither home nor office – an alternative place where work can happen. This has been bubbling up with cafés acquiring a new role and becoming a possible work environment for the 'digital nomads' who wander through their neighborhood 'looking for a place to call work.' Now we are seeing co-working spaces that house lots of freelancers and start-ups in open environments set up for 'pay as you go' working. Such ventures as General Assembly have attracted venture capital investment from Silicon Valley, evidence that there is money there and that this is a trend with a strong future.

Another third place trend started to take shape in early 2012, when Shell and serviced office company Regus piloted a new venture for 'drop-in' offices at motorway service stations near Paris.

Case Study

The Open Work program at Sun

The Open Work program at Sun Microsystems was one of the most sustained and significant early initiatives to take an integrated view of technology, office facilities, working practices and HR policies.[17]

Led by Ann Bamesberger, then VP of the Open Work Services Group, the program had a massive impact on changing working patterns, increasing employee engagement, reducing the firm's carbon footprint and slashing business costs. Sun also used its experiences to develop an offering to clients who were looking to reap similar benefits.

Sun was genuinely a leader in developing an holistic view of the Digital Workplace, through a program that saw itself as 'an integrated network of people, places, and technologies that are systemically linked to meet the requirements of the task at hand.'

Beginnings and philosophy

The Open Work program started as early as 1994. Initially, the Open Work Services Group set out to look at office design but, early on in the process, the group forged important partnerships with Sun's HR and IT functions, which allowed them to incorporate technology and working practices. The program was a natural extension of Sun's corporate philosophy that the 'network was the computer' and that 'everyone and everything was connected to the network.' From this, the concept of working anywhere and on any device by connecting to the network was easily derived.

Working patterns

The program was about 'an expansion of choices for work style.' Employees entering it were assigned either to be home workers (three or more days at home) or flexible workers with no fixed location, generally working from home two days or fewer. Others could choose to remain assigned to a seat. Depending on their status, they were given significant technical support to ensure that they had the necessary equipment and set-up in order to work successfully. They also had privileges to reserve workstations in the variety of working locations on offer.

The Physical Workplace

Sun provided a variety of office locations in which to work. Traditional Sun office locations had flexible 'zones' and Open Work cafés with workstations, some of which could be reserved in advance. They also provided dedicated 'drop-in centres,' which were located in major cities, on Sun campuses and even on commuter routes. The idea was that employees could work there in between meeting clients or to avoid commuting at the worst times.

With around 70% of the workforce mobile to some degree, it was not surprising that Sun was able to significantly reduce its office space. In fact, its own research in the early 2000s had shown that, regardless of the flexible working program, 35% of the workforce assigned to an office were not present there anyway.

The technology

At the centre of the program were 'Java Cards' which doubled as security cards. These could be slotted into a workstation and within seconds brought

up the employee's personal desktop, perfectly preserved from their last session. Thanks to this Sun Ray technology, employees didn't need to carry laptops around at all; they just plugged into the network. Moreover, many employees had this capability at home.

Cost savings and environmental impact

Sun estimated that it has saved US$70 million in real estate costs and US$24 million in IT costs every year. Sun employees could correspondingly save up to $2000 a year in fuel costs, as well as reducing commuting time by up to 160 hours a year.The overall environmental impact of this, mainly from cutting down on commuting, was an estimated reduction of 52,000 metric tonnes of CO_2 a year.[18]

Support and training

The program also invested significantly in support tools and training. There was a dedicated area on the portal, and a tool called 'Open Work Select,' which was an online assessment tool to allow individuals and managers to work out employees' suitability for flexible working. SunReserve allowed you to book workstations. A 'distance collaboration' program not only ensured that meeting rooms were properly equipped, but also provided training to run virtual meetings. For example, one strong piece of best practice was that virtual attendees are often at a disadvantage and can get forgotten about if half the meeting is physical; it's much better to have the whole meeting virtual. There was additional training on how to manage remote workers.

Cultural aspects

Although there was inevitably some opposition from managers, one very noticeable feature of the program was that it was very respectful of the existing office culture and didn't necessarily lead through a big change management project. Bamesberger commented: *We've not forced anything, so we've learned to really very, very carefully enable and watch for next-generation behaviors.* Instead, support was given through elements like the suitability assessment tool.

Impact on time and productivity

Time saved through the program was regarded as a real win–win. *Our self-perceptions of our increased productivity, due to being able to make choices on the margin, have really, really come into play here. I think people are seeing the advantage of not having to come in for unnecessary trips, and the absolute value of coming in for those trips that are worthwhile. And I think that the increased productivity from not having to make non-essential trips is given back to the company and is also taken back by the employee as a benefit,* said Bamesberger. She estimates that 60% of saved time was devoted to work and 40% to family time.

Chapter 9

Workforce

The new workforce will be older and younger – and thinner!

What we are seeing in the major new technology giants of Google, Facebook and Twitter is the ability to produce massive value with few employees.

How will the Digital Workplace impact the workforce? For centuries, mechanization has affected the skills, location and demographics of those in work. Farming was revolutionized by the introduction of machinery; the woollen mills were reconfigured through automation; and the story of the Ford Motor Company and its production line systems defined the last century. Technology has turned factories into sparsely populated locations where a tiny number of people produce extraordinary levels of output. So how will the development of a digital world of work reshape work for this century?

It is impossible to predict exactly what will happen but certain trends, which will accelerate, can already be seen. What is fascinating is the way in which the ability to work digitally is not only creating a new skill set and 'pattern' of work, defined by flexibility, projects and adaptability, but that it is also redefining *who* can and cannot work. Rather than requiring retraining and reskilling, the effects that can already be witnessed are changing the age profile and workforce constituency. If this is rolled forward in time, the extent of this workforce change will leave us breathless.

Gone abroad

Let's talk about actual experiences. For the past two decades, work has been migrating from the advanced and expensive western economies to lower-cost, developing nations. Moving work 'offshore,' as this has come to be known, has taken roles, processes and production to India, Eastern Europe and the Far East. It is not so much the obvious 'we gave all our manufacturing to China' syndrome (which in any case is not a product of the Digital Workplace but a consequence of low wages and global shipping), but the way in which professional services, information management, customer services and IT have been relocated to countries like India, Thailand and Mexico.

The Digital Workplace has already been a major catalyst behind the explosion in outsourcing as work has shifted location. When work becomes released from physical constraints, it can move to wherever it can be completed most cheaply and quickly, so long as quality is maintained. If you take a major corporation today, such as Shell or Toyota, their work is spread far wider geographically than it was 20 years ago. Tasks, processes and systems can be broken into elements and placed wherever they can be most efficiently achieved.

'Work stretching'

One key trend that is emerging as work reshapes itself, enabled by digitization, is the ability to have older workers remain productive for longer and longer. Staff leave their organizations when retirement becomes an option but then have the possibility of remaining engaged and delivering value for years to come.

One former middle manager I spoke to recently spent his working life at a major energy company and now works 60 days a year for them as a consultant, writing and delivering online learning for staff across the organization. He has a physical meeting once a year with the employee he reports to and works from a pleasant base in France. He devotes a similar number of days a year to helping a charity which builds homes in areas struck by natural disasters.

What we are already seeing, and will increasingly see, is that the Digital Workplace is allowing what I call 'work stretching' as people find they are healthy enough and motivated enough to remain productive well into their 70s and 80s. Companies who need their skills are happy to pay reliable, self-motivated and cost-effective contractors to produce for them, and this workforce age profile will gradually extend further into what we currently call 'old age.' With the number of currently young people now expected to live beyond 100 years, the whole concept we have around what 'old' means will be revolutionized as a vibrant, technically adept

The Digital Workplace is allowing 'work stretching' as people find they are healthy enough and motivated enough to remain productive well into their 70s and 80s.

and efficient workforce in their 70s and 80s becomes normal.

Given the current justified crisis around providing pensions for an older and older population, this new facility for post-retirement working is a boon for governments, as well as for citizens. Cut out the soul-destroying, not to mention body-destroying, commute to work and older people can be liberated by new careers that start at retirement rather than end there.

Companies will start to 'recruit' teenagers to work for them before they have even left high school.

Child labor?

But that is just one end of the age spectrum. A maybe more startling – and certainly more challenging – idea for us is what might happen at the other end of the age spectrum. Recently, a 12-year-old American, Thomas Suarez, spoke at the celebrated TED conference about the app business he has developed and commercialized using his high levels of coding and technical skills.[19] The issue we face is that some of the key skills that organizations require in technology, new media and online services are held in the heads of younger and younger people. Power is shifting down the age range as it dawns on the likes of Thomas Suarez that they have access to everything they need in order to create new, profitable services. This is a remarkable change.

My prediction is that 'work stretching' will shoot down the age range as many companies begin to 'recruit' teenagers to work for them before they even leave high school. Precedents for this exist already in the fields of sport and the arts, where young, gifted children are trained within organizations and industries, acquiring value for the organizations that invest in them. For example, in the UK, soccer academies contract with children at 12 years old, and even younger, to make sure they have the best new talent at 16 and 17 years old. How long wil it be before Google, Microsoft, and even the more traditional companies, start to bring teenagers onto the payroll because those youngsters have something to sell and their parents perceive this as a positive rather than a negative experience for their children?

Will schools be able to compete when Google offers the likes of Thomas Suarez US$100,000 a year to join a specially constructed 'Student Staffer' program? Gifted children would be able to study at a 'Google School' combining the best of a conventional education with a job delivering new products for the employer. This might seem like some horrific new-technology version of a return to child labor but, in this scenario, both the child and family would be well remunerated for the work, while their schooling would be far from neglected, in fact coud be specifically tailored to the high-performing student.

The impact of the Digital Workplace is profound and will create both positive and negative impacts. I am not saying this is universally a great development but just that it is coming to a company and home near you.

'Work stretching' will see a workforce starting work at age 13 and ending at age 90.

And the magic diet?

So that's the older and the younger. The bit you really wanted to hear about was how the Digital Workplace can make you 'thinner'? Well, sorry, it's not you who is going to get thinner but the workforce itself. What we are seeing in the major new technology giants of Google, Facebook and Twitter is that you can produce massive value with very few employees. These companies – even Microsoft, which has been around a while longer than Google – has a relatively small staff. Google has 20,000 as I write, Microsoft 87,000 and Twitter just 650. Compare that with Toyota employing 320,000 or Deutsche Post DHL with some 500,000.

But the model used by the likes of Google will be implemented more and more across all organizations as technology removes human beings from work. So, what we will increasingly see is 'thinner' workforces, with major corporates employing a leaner workforce and the only expansion happening at the younger and older ends of the age range.

This does present a major challenge around the human population and jobs. What will we do if the jobs required diminish due to digitization? New industries, new services, new markets? Yes, they will emerge but the challenge for governments may be extreme.

Chapter 10

Recruitment

How to lose your best new hires – fast!

It can seem so fabulous. On the university recruitment circuit, and in the brochures and You Tube-style 'what it's like to work here' pieces-to-camera from recent hires, they seem like such great people. You get the job and, excited, arrive for work at your new high-profile employers.

Here's your desk (slight sinking feeling) and there are your colleagues (they seem friendly and smart, but they're all busy) and then you log on with your newly issued password. The system fires up and the latest incarnation of Windows powers into action, surrounded by all the standard icons. You activate the browser to reveal a dull, old-fashioned default intranet home page. Next you hunt for the employee directory (becoming a bit panicked now) and eventually locate a pre-Facebook and Twitter world – well organized but very boring. Increasingly desperate, you begin to click around the screen real estate like a traveller in a desert seeking water. Finally, you resort to Linkedin to check out the profiles of staff from your new employers… Access denied.

New hires are like water. If you don't give them what they need in the digital world, they will flow into places that do.

Great firms are losing new hires because their Digital Workplace is so poor. All that money spent on presenting a great face to new recruits counts for zero when the technical reality hits home. I have spoken to HR folks at five major corporates in the last two months who all said they knew of recent hires who had moved on partly because of the shocking state and interface of the work technology they found there when they arrived. This is much more than just a cosmetic problem. New recruits expect the best designed, most connected workplace possible; one that will allow them to work to their maximum. If not, they go back to Facebook and moan to their friends who compare and contrast.

What do you mean, there's no Facebook?

If you are going to an interview and, ahead of it, you Google the company only to learn that Facebook is 'banned,' do you care? In terms of being able to access Facebook at work or not, you are probably not too fussed but, more importantly, what does this exclusion tell you about the working environment? It suggests an inward rather than outward looking stance and a mistrust of the staff.

I recall one middle manager at a major telecommunications giant who said that people arrive at the company with high hopes but, after a while, the corporate culture just grinds them down. He seemed faintly proud of this, maybe because it echoed the compromises he had himself made.

But while the Digital Workplace affects so many parts of organizational life, with new recruits it matters deeply as you will lose staff – or never acquire them in the first place – unless your Digital Workplace delivers as well as the external services 'Gen Y' people are used to.

This ability to 'compare and contrast' didn't exist before. New hires in my day (1979 remember!) had nothing to compare the work environment they experienced on arrival with. Now, new hires can fire up the web browser and make instant assessments of the intranet and any related services they find – because they already know what good looks like. They have experienced thousands of websites and are constantly using social networks. Generally, the results when they arrive at work are not great and work depression starts to set in fast.

And, in the online Gen Y world, reputation spreads fast. 'Best companies to work for' will very quickly lose that status if their Digital Workplace falls flat on its face. It amazes me that organizations, both large and small, don't realize how important a high-performance Digital Workplace is for people who start work there.

At IBF we delight in demonstrating the technology we regularly make use of in our daily work. At interview we show off the 20-plus online services we utilize constantly – Basecamp, Google Docs, Skype, WebEx, Live Meeting, InterCall, extranet, website, four blogs, three micro-blogging communities, Linkedin, Twitter and

VPN – and people get very excited at the range of technology on tap.

'Freelancing' the workplace – your global resource

Since her first day with IBF, Helen Day, now our Global Managing Director, has worked from home in Nottingham. She started life in IBF as a freelancer and flexible working is crucial to Helen. When, after two years, she came on staff and proceeded to 'rise through the ranks,' she held firmly to her preferred option of working flexibly from home. Her example has helped to shape the results-driven style of working in IBF and Helen was a key driver in helping me have faith that removing our last physical office would be fine.

In the Netherlands, **1 in 2 highly educated employees work from home;** among workers with a secondary education it is 1 in 5; and for the less educated 1 in 8[20]

When Faye Andrews, our Marketing and Communications Manager, resigned to start her own business, she left her staff role but continues to work for us as a freelancer, looking after the events side of IBF, which had been her staff role.

Mark Tilbury, formerly Intranet Manager at BDO Stoy Hayward was once the key representative with BDO for their membership of IBF. Then he became IBF 24 content producer, working for us freelance, before moving to KPMG and again being a customer.

In each of these instances the structure and nature of the relationship changes but it all happens within the unchanging (although evolving) Digital Workplace. People shift conversations and contacts but there is an easy fluidity that comes because the relationship is always digital.

Opening up the world

The Digital Workplace unleashes a whole set of new flexible ways of working and connecting commercially, and also opens you up to the world's freelance resource pool. Anyone can work for you from anywhere – and, likewise, as a freelancer you can work for virtually anyone, anywhere.

The proliferation of marketplace websites such as eLance, oDesk and Freelancer.com is testimony to the new muscle of small businesses in their ability to hire professionals, almost immediately, with next to no overheads, taking advantage of access to a much bigger pool of workers. Larger companies are also getting in on the act.

They may use a 'crowdsourcing' company who will take a very large project, break it down into tiny tasks and feed these to a registered army of thousands of online workers. Some of these companies use their own in-house developed platform to connect; others just dish out the tasks via Facebook.

The restriction of needing to recruit locally is removed and the problem of 'so and so calling in to say he can't come in because of the train strike' just goes away. For freelancers and employers this is a liberating experience. Physical limits are replaced with digital limits, except that few limitations exist in the digital world, while in the physical world limits are everywhere.

The Digital Workplace unleashes a whole set of new flexible ways of working and connecting commercially, and also opens you up to the world's freelance resource pool.

Do's and Don'ts

- Make sure your Digital Workplace meets the expectations of your Gen Y new hires – it's going to be an increasingly important part of your employment proposition.
- Think twice before you ban access to social networks. What does it say to the workforce about your company culture?
- Take advantage of the world of freelancers. There are some brilliant people out there who can come in and out of your organization and will continue their relationship with you in various different ways.
- Your digital footprint on the web will be considered by prospective employers and employees. A certain lack of privacy is now culturally accepted.

Case Study

Leveraging the global talent pool: Gary Swart and oDesk

The rise of the Digital Workplace has meant there are new opportunities for freelancers. This has seen the rise of service marketplace platforms like oDesk and Freelancer.com, which match employers to contractors, and then use technology to facilitate virtual working.

In fact, oDesk describes itself as 'the world's largest, most comprehensive and fastest-growing online workplace.' In 2011 alone, they claim to have posted over a million jobs, with more than US$225 million earned by contractors.[21]

There are several conditions that have allowed companies like oDesk to flourish. The company's CEO, Gary Swart, believes there are three main global 'mega-trends or enablers': the economy, globalization and technology.[22]

The first enabler is the economy. The economy on a global scale is forcing companies to look for better ways to do things, so companies are trying to do more with less. The second trend fuelling the growth in this new way to work is globalization. Many of our jobs are onshore, meaning US employers and offshore contractors; however, we're also seeing massive growth in offshore-to-offshore and offshore-to-onshore working relationships.

The third mega-trend is the Internet and technology. The technology enabling you and me to communicate and collaborate just continually gets better and

better. With tools like Skype, it's as if you're sitting next to me, and with platforms like oDesk, you and I can work together as if we're in the same office.

Although the job-matching of oDesk is important, their USP is the technology that enables projects to be managed effectively. Contractors can bid for new opportunities, manage their time, get paid, message their employer and connect with other members of their virtual team. Management can have a dashboard of time billed of all the contractors in their 'team room.'

Employers can also check that contractors are actually working during any hour of billed time as, via the oDesk application, six randomly timed screenshots of the contractor's desktop are taken in any one hour and then sent to the employer. If the contractor lets this happen, then payment for the time worked is guaranteed.

oDesk have used this model for their own growth: *We have 75 employees, and roughly 300 contractors that work for us every day, from around the world. We eat our own cooking. We built our business leveraging using our own network, and we still use our own network today, almost to the tune of 4:1.*

Swart believes that the market is only going to grow. *I don't even think we're near the knee of the curve of what's to come, as larger companies start to say, 'Hey, wait a second, what about us? We need to save costs and we'd like to have fewer bodies and less office space and infrastructure for these employees. Is there a way we can leverage a global talent pool in order to get work done?'*

Case Study

Flexible working at BMW – and the level playing field

BMW has been a pioneer of flexible working patterns since the mid-1980s. Originally the impetus for introducing flexible working was to increase production capability at its plants, but today this is much more about individual employee satisfaction and engagement. The extension of these benefits to all employees – creating a 'level playing field' – has also been a catalyst for the development of the Digital Workplace.

The first flexible working experiments were made at the firm's Regensberg plant in Germany as early as 1986. Shift patterns were introduced that concentrated shifts over four days rather than five, involving some Saturday work but allowing for significantly longer five-day breaks between shifts. Production was subsequently increased, partly because the factory now ran for six days rather than five.[23]

To compensate for fluctuations in customer demand, some of which were seasonal, the company also introduced the concept of 'work time accounts.' A worker could accrue or owe up to 200 hours (later extended to 300 hours) in working time that could extend over years, allowing for even greater flexibility. The work time accounts were protected by agreements that covered payment for overtime and individual arrangements, so that the needs for the individual and the company were theoretically in balance.

Through the 1990s, the firm continued to establish a number of local flexible working agreements within its structure. At one point, there were over 300 different schemes in operation, including a special scheme to create

temporary part-time positions for workers; this is still in operation today under the name 'Full Time Select.' Sabbaticals were also encouraged.[23,24]

The flexible working policies at BMW have helped stimulate the development of a Digital Workplace. In order to create a level playing field, the flexible working policies were extended to nearly 20,000 office-based staff (BMW in Germany alone employs around 70,000 people, of whom around 40,000 are on the production line.)[24] In the early to mid-1990s the firm also implemented some early experiments in teleworking,[25] which were subsequently extended. Flexible working became a key part of BMW's employment proposition and people could happily work from a number of locations.

However, as more core processes within BMW were put online, this effectively created a 'digital divide' between the white collar and blue collar workers. For example, applying for the special benefits offered to employees relating to BMW vehicles was organized through the intranet – but the majority of staff did not have access to a terminal during the day.

Again, to create equal opportunities for all workers, accessing core systems and the intranet from home was encouraged. In addition, teleworking became easier through a series of initiatives, for example, allowing private network access by 2007.[24] Overall, allowing home access to the company's systems became an important method of improving the flexibility of both associates and the company. Today more than 20,000 associates are able to work remotely.

Top ten ingredients for a successful Digital Workplace

1. CEO and executive level support – and, ideally, practice.
2. Middle management training in managing in the Digital Workplace.
3. IT function that enables rather than prevents portability and access.
4. Shift from input-focused work to an emphasis on results and outputs.
5. Proper internal communication and engagement ahead of the changes.

6. Measurement and metrics around the impact of the Digital Workplace.
7. Clear strategic vision of the shape of work for the organization for the next three to five years.
8. Visualization for senior management (in pictures and video) of what a typical day will look like in the Digital Workplace.
9. Promote and communicate 'quick wins.'
10. Technology that works smoothly and efficiently.

Close

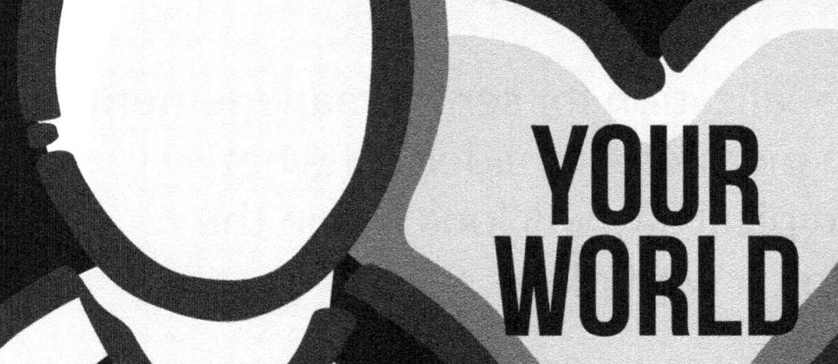
YOUR
WORLD

Chapter 11

Isolation

I feel a bit lonely!

People often meet their partners at work. They make mates at work and go for a drink after work with their colleagues. After a stressful weekend with the kids, their colleagues help take their mind off it. Work is therapy for life. At least some work is, some of the time.

The point is that work in the Physical Workplace has a social, communal and human dimension to it, that is really nothing to do with the work itself; it is about our basic need for contact and community. I recently met a former (young) colleague who now works at a big television station and she said that, generally, she really enjoys dragging herself through the underground system into the office each day because when she gets there it's a buzzy, social and interesting place. Yes, home working now and again would be good, but in the main she likes the feeling of being a part of the media industry.

> It's not black and white – either 'full contact' Physical Workplace or 'zero contact' Digital Workplace – it's about shaping work to suit your agenda.

The younger you are, the more likely this is to be the case. The Physical Workplace offers structure, connection and discipline. Also, the younger you are, the more likely you are to live somewhere that is not suited to home working, or perhaps you still live with mum and dad and they're already working there. Getting used to an office helps with your training and education in the world of work.

But fast-forward a few years and the 'thrill' of going into that wonderful office, day in day out, starts to look a bit threadbare: the same commute, same faces, same desk, with all the inflexibility that entails. Yes, you can arrive a bit late and leave a bit late to miss the rush-hour but after a few years that doesn't really cut it.

In IBF, the staff and freelancers who have experienced a few years in an office, view the Digital Workplace as a liberation. As mentioned earlier, Angela Pohl, now IBF Managing Director, who is based in Ohio, had a job in a large pharmaceutical company before leaving to have kids. She returned to work through IBF and adores the Digital Workplace as it gives her all she wants from work – she feels productive, well compensated, connected and intellectually stimulated. She gets to meet her colleagues several times a year at member meetings in the US, and on trips to the UK every year or two she meets others she has only previously met virtually. It is perfect for her.

At IBM, the catchphrase used by some staff is I – B – M, standing for 'I'm By Myself' and many large organizations experience the same issue. That said, we need to adjust. It's not black and white – either 'full contact' Physical Workplace or 'zero contact' Digital Workplace – it's about shaping work to suit your agenda. If you want to collect your kids from school, or do the shopping during the day, or just prefer to follow your own body clock and colleagues' working hours, then you can organize this to suit you, within reason. In IBF, we have specific days when the team work physically together in London and New York, other days when we generally work from home and then days when people flex as it suits them.

For me, the potential loneliness of being 'stuck at home' in your back bedroom office is about feeling that this is the only option. It can become as tedious as commuting into the same office each day, although research suggests that most people enjoy the upside gains of home working more than they do the office work experience. It is important to shape your week, month, year so that you have variety, ideally with some days at home, but don't lose sight of the alternatives – meeting colleagues to work with them in a shared office space, cafés, each others' homes, customer venues and so on – there are many options out there.

When I first started working from home (many years ago) people always asked 'How do you discipline yourself?' It's an adjustment. Work requires certain levels of discipline. If you stay at home and do nothing much, then the results of that inaction will surface in your work and there will be consequences. If the working pattern doesn't suit you, then change it to one that does. The Digital Workplace is not about home working; it is about working where you want, when you want – and sometimes that might be in the conventional office.

American Electric Power: Photo Archive

AEP uses its intranet to overcome any isolation people may feel through a simple device. The company has a long history and it asked people, many of whom had been in AEP for several decades, to post archive photos relating to the AEP history and to ask others to provide some historical context. This has been hugely successful, quite moving at times and basically free to do.

So where do I meet my life partner?

So, if people like making friends at work, and maybe even meeting a romantic partner too, where will this happen in the Digital Workplace? Well, we could say that it's hardly the role of the workplace to act as a dating agency but, that said, the social side of work does not just evaporate in the Digital Workplace. In IBF, many people are friends with each other outside of work – no lifetime connections as yet, but we are still fairly small! People form connections virtually and then use the various scheduled physical workdays, customer connections and random hook-ups to deepen those connections. By the same token, with the success of Internet dating riding high, there's no particular reason why the socializing of staff in digital spaces might not be just as likely to throw up the occasional romantic interlude as when colleagues meet in person!

Large companies with high levels of remote working need to think holistically about how people working there can feel better connected with their employer and to the colleagues they work with directly. Within IBM, there are large numbers of staff who now have a decade of working outside of a traditional office and the only problem cited is one of isolation. This affects some IBM staff more than others and can be countered by ensuring regular meetings in person and the days in the offices IBM still has.

Managers need to schedule 'face time' with their direct reports for productive work, not just to 'say hi.' They need to accept the impact of potential isolation both culturally and socially. Closing down swathes of offices is fine but these need to be replaced with 'hubs' where local colleagues can work if they want, enabling physical meetings to happen in order to offset the separation effect. We are social – but as Facebook shows, the social network can be very virtual!

With the success of Internet dating, there's no particular reason why the socializing of staff in digital spaces might not be just as likely to throw up the occasional romantic interlude as when colleagues meet in person!

Chapter 12

Addiction

Confessions of a connection addict

One of the potential issues for those involved in the Digital Workplace is addiction. Addiction is a strong word implying physical and psychological dependency. Here we're not talking about being addicted to the Digital Workplace, or even to the process of working virtually – it is more a question of addiction to 'being connected' to the Digital Workplace.

A recent study from Cisco found that half the teenagers they spoke to regarded being connected as important as having water and food.[26] Here's an example from my own recent experience...

On my first day of holiday I left my iPhone on the plane. Panic and stress lasted for a few hours but then I just made sure those who mattered knew what had happened, set up an alternative number for emergencies and relaxed for a lovely week. Not having the iPhone was fine.

But that was on holiday. When I arrived back home ready for work, it was a couple of days before a replacement phone could be set up. It was awful. Not only could I not work practically, but I also felt 'unconnected' and impotent in both work and energy. I felt I might as well not even get out of bed as I was unable to access the Digital Workplace at all. The loss of connection lasted just two days but those 48 hours are etched in my mind. Ladies and gentlemen, I am an addict; not recovering, still at the peak of my addiction and, if anything, it's getting worse through the Digital Workplace.

I sit here writing on my Mac with my iPhone positioned beside me: signal strength solid, texts popping up here and there, email at the ready, Yammer on the go, Skype present and correct, Twitter ready to tweet, WhatsApp available, and on and on… that makes me calm and happy, like a man supplied with his drugs of choice. Good, bad, indifferent – not sure – but in any addiction, Step One is to admit the problem. Trouble is I'm not sure how much I want to stop the habit!

In fact, we are touching on two significant aspects here. As technology improves, being connected anywhere, anytime becomes increasingly normal. Alongside this, we find that the Digital Workplace with the power it gives us makes work more enjoyable and satisfying. So what we will see is a new 'health epidemic' due to work addiction. More and more people, both freelance and employed, will find themselves working more and more hours, because they can and because they love the work. Weekends, evenings, holidays will all blend into work as 'rest time' decreases. Work addiction will capture media attention and cause healthcare

'Alone Together'
by Sherry Turkle

Turkle is like Jaron Lanier for me (author of *You Are Not a Gadget*) in that she reveals the dark currents running through our supposedly connected lives. She talks about the hook of always being online and accessible by phone. We feel as if we are 'together' but we are actually alone. We update our status and check for messages when we could be talking to the person beside us. Drugs come in many forms.

concerns in much the same way obesity has. It is ironic that one of the reasons for this will be greater work satisfaction and fulfilment. The problem is that if we can work whenever we want *and* we enjoy the work we do, we will work more and more hours, until gradually every aspect of our day becomes colonized by work. Work addiction detox problems will emerge, the media will lap the issue up because they can relate to the issue themselves and employers will have to take this seriously (while all the time quietly enjoying the productivity benefits of their 'always on/always there' staff).

> Yes, you can work when you want but colleagues and bosses will also feel free to connect with you outside normal working hours.

Why employers secretly feed our addiction

As work becomes present anywhere, anytime and also more enjoyable due to the new freedom technology allows, there is a big danger of our addiction to our 'drug of choice' – work – getting stronger. Might employers secretly feed this habit to squeeze more and more out of us by making sure we get more of what we crave: beautiful devices with ever better services, always there, even resting – tempting– beside our pillow while we sleep?

On holiday in recent years the day would pan out like this... Breakfast on the terrace before people moved to the pool for relaxation and sun. While this was happening, the adults with engaging or perhaps purely demanding jobs (just about everyone in work in fact!) would check their emails via laptops and smart phones, make a couple of calls, send a few messages… 'just to clear the decks'. Later, you would bump into them again, by now actually relaxing and holidaying with their families. But the whole day would be punctuated by their steadily and persistently checking emails, the phone and texts 'just to make sure.'

When pressed about this habit, people get defensive, play down the tendency, explain how it allows them to actually take holidays, and other excuses. The truth is that many people just can't – and actually don't want to – switch off from work. They have become addicted to staying connected. Of course, this quietly thrills employers: more work for the same money!

It is one of the counterpoints to the Digital Workplace that the liberation which results from being able to work when you want, and increasingly *where* you want, is also feeding an addiction to staying 'in touch.' Yes, you can work when you want but colleagues and bosses will also feel free to connect with you outside normal working hours.

I recently gathered a few middle managers together in a bar in Chicago and we chatted about the Digital Workplace. They seemed to consider that working 'all hours' was fine because the payback was so valuable. Now, that's their choice of course, but when I suggested that weekends and evenings should be leisure time, the reply was, 'Yes, but… .' Isn't that the kind of thing people say when they're accused of being an addict?

Possibly because I feel as if I have been working in the Digital Workplace for 25 years now, the boundaries for me are clear. I don't work evenings or weekends, and on holiday I check emails and the phone once a day at 6 p.m. – but expect only to hear if there is an emergency. As CEO, the 6 p.m. check-in seems a sensible risk management calculation. I don't call people in IBF outside normal hours with the odd exception and I don't expect to be contacted at those times either.

The implications of the Digital Workplace seeping into every minute of our lives will have a negative effect over time. But perhaps the benefits of the Digital Workplace are so great that this 'addiction' downside may prove to be worth the pain. It is always true that when big shifts occur there are conflicting forces at work. Let's see how this aspect pans out but I would advise anyone to set clear markers between work and non-work; we need that separation as human beings.

According to the Telework Research Network,

at least 45% of the US workforce hold a job 'compatible with at least part-time telework.'

They believe there are 50 million employees who have telework compatible jobs and want to work from home. If those who had compatible part-time telework jobs worked on average 2.4 days per week from home, the national savings per year for the US would be an estimated US$900 billion[27]

Employers are in an interesting place with this. They secretly like the increased availability of staff and their flexibility, and can exploit this practice. In which case, we may see legislation emerging to attempt to put some health and safety policy in place – which sounds like a disaster in the making given the bureaucracy involved. My belief is that work addiction through technology will become a new health epidemic as powerful in its 'media profile' as obesity is today; an unnoticed new society-wide problem where work pervades every part of our lives and non-work time dwindles in size and space. Separating work and leisure remains critical – but try telling that to a work addict!

My belief is that work addiction through technology will become a new health epidemic as powerful in its 'media profile' as obesity is today.

Are BlackBerries really addictive?

The BlackBerry was, of course, the first mobile device to really penetrate the corporate world. With it grew the reputation of the 'crackberry' – recognition that this truly game-changing technology was addictive. There was widespread media coverage about the dangers of becoming hooked, and how the world of work was seeping into the world of non-work. But are BlackBerries and similar devices really addictive and potentially dangerous?

Different academic studies in the past few years have tended to suggest that BlackBerries *are* addictive. Research from Rutgers University has equated BlackBerry addiction to drug use, and suggested that firms need to be wary of impending lawsuits.[28]

In 2008, an Australian study conducted nearly 30 interviews with senior bankers. Dr Kristine Dery, from the University of Sydney concluded: *There is a real problem for organisations where stress, burn-out and addiction to 'crackberries' are real threats to long-term talent retention and organisational effectiveness.*[29]

Furthermore, tests carried out by UK-based psychologist Dr Glenn Wilson suggest that the constant 'change of direction' in the brain triggered by the compulsion to reply to an incoming email poses a greater risk to levels of concentration than smoking cannabis.[30]

One of the most comprehensive studies was by MIT Sloan School of Management in 2006.[31] They spent several months observing the behaviour of staff using BlackBerries at a US private equity firm (referred to by the fictitious name Plymouth Investments). The study found that individuals

were sustaining 'an almost constant connection with their organizational lives.' The researchers observed what they called three 'dualities' – contradictions between the perceived benefits and the actual resultant behaviour amongst BlackBerry users.

The first of these was between users' continuous observations of their business dealings on the BlackBerry, yet their responses to emails were 'batched' together and were anything but continuous. The second was between their engagement with the world of the BlackBerry and their corresponding withdrawal from the social world around them, causing tension with family members. The third was between the autonomy they felt by having control over time, tempered by the addiction to the device.

This shifted expectations of availability and responsiveness, generating increased dependence on staying connected, and a compulsion to constantly monitor the flow of email communication… Peer pressure, firm norms, and professional identity encouraged members to constantly check and use their BlackBerry device. Such constant checking led to a certain compulsion (even addiction) to being in the flow, a connection that further fuelled expectations and dependence.

This potential 'addiction' to 'being in the flow' was brilliantly expressed in some illuminating comments from the partners of those working at the company. One woman observed:

Well in some ways checking your BlackBerry is like pulling the lever of a slot machine… Maybe there'll be an email from someone you haven't talked to in a while. Maybe there'll be a joke or maybe there'll be a good piece of business news. I mean, I think there is that sort of sense of anticipation and potential gain that you get from checking. I think that's the addictive part.

Another observed:

I think it's more a matter of feeling that he's indispensable – and needed. Somebody somewhere needs his opinion, or needs a conversation, or cares to engage him... I think that they're addicted to the idea that someone needs them all the time.

And this behaviour is continually reinforced. One member of staff observed:

Once the audience that you interface with all the time know that you're a crack junky, then if you don't respond to an email in an hour people start to wonder 'What's wrong with Gary?' I mean it's that bad.

Chapter 13

Work satisfaction

Work is good… but it isn't everything

The Digital Workplace does not create work you love. I have already alluded to the tedium and pain of most work in most organizations and how a large part of most people's work is actually unpleasant to them. Learning to love your work can result from a range of factors and, while it may be a place to aspire to and a blessing when you discover it, working in a Digital Workplace will not automatically turn a dull job into a wonderful job. But it will improve the conditions in which you work and will give you more control and power over how and where you work – and these are key factors that impact on the enjoyment and even love of what we do.

The Digital Workplace changes your relationship with work so that even a possibly routine job, like working in a call centre, can become more rewarding if you can do it when you want and from your home, or while visiting friends overseas, or staying with your mum and dad. It enables choice and that makes us feel happier.

"I say to my players, what is the hardest thing to achieve in life? I think one of the hardest things you can do is to work hard all your life. I don't think that's actually easy. Working hard is actually good for you. I get it across to my players that working hard is a quality."

Sir Alex Ferguson, Manager, Manchester United Football Club

Do I have a life left?

So, we have looked at the flexibility you gain with the Digital Workplace, but what about the effect on those you live with? If you choose to work from home as part of the Digital Workplace, for example, what should your family do when the kitchen becomes your temporary Physical Workplace? The short answer might be to organize a room or place that is less intrusive to the others in your home. But the bigger issue is around the work/life balance. The Digital Workplace blurs the boundaries of work and life and, strangely, most people I talk to seem to like that blurring. They say they enjoy, or at least accept, the fact that work will seep into their 'non-work' time.

> Boundaries between work and life are important. In the years during which I worked in our IBF office (actually a garden cabin) about 10 yards from my house, this short distance created a physical separation that was perfect.

To me, this is a big negative. It is unhealthy and unnecessary. It has come about through creeping work addiction or, more accurately, 'connection addiction' on the part of staff, managers and colleagues alike who believe that the Digital Workplace somehow makes it acceptable to Skype someone at 8 p.m. or 7 a.m.

Boundaries between work and life are important. In the years during which I worked in our IBF office (actually a garden cabin) about 10 yards from my house, this short distance created a physical separation that was perfect. The tiny 'commute' was fun and a regular joke with my family; at the end of the working day, I could lock the door and 'go home'!

Must we be 'crazy busy' all the time?

What is happening is that the speed of technology is changing the way we approach time. The Digital Workplace plays into this – but only to the extent you allow it to. We have come to expect an immediate email confirmation if we book tickets for travel or the theatre online; we post a question to colleagues through Yammer and assume there will be instant replies. The nature of action and response has become accelerated.

So what effect does the Digital Workplace have on time? The boundary between time working and time not working, and the blurring of this is being experienced more and more. This is to be avoided. We can control the time collapse from the Digital Workplace but it appears hard to manage due to shifting work patterns and the fact that employers rather like people working longer hours (although they will never admit this).

We are seeing an acceleration in time; that's a given. However, there is a difference between being 'crazy busy,' which is a negative all round, and being fast and responsive. I have always liked Stephen R. Covey's action quadrants[32] around dealing first with what is 'important and urgent,' then with 'important and not urgent' and relegating 'urgent but not important,' which occupies a lot of work time, to last. The really significant point though is that 'not important and not urgent,' an area many people in many organizations devote huge amounts of pointless time to – because they actually enjoy it – can in fact be ditched altogether!

Chapter 14

Privacy

No such thing as private

The digital world may give us an insight into the organizations we work for, or are applying to work for, in rich detail. But this cuts both ways. Your potential employer, as well as your current employer, can learn a great deal about you before they even see your face in person or on Skype. While you are on the payroll they can track and observe you if they so wish. Sure, you check out their true colours – but so they can yours too. So let's say you are not visible on the web; you are private and happy with this. But if a prospective employer searches for you and draws a blank, what does that suggest? That you're secretive? Private? Technically challenged? Or just odd!

If a prospective employer searches for you and draws a blank, what does that suggest? That you're secretive? Private? Technically challenged? Or just odd!

In fact, privacy is becoming, for better or for worse, a relic of another era. New hires accept that you will have found out all you can about them and recognize that you will expect them to have found out as much as possible about the company.

What is privacy now?

So what is the nature of privacy in the Digital Workplace? Is it any different to the Physical

Workplace? When Nokia started using GPS for staff whereabouts, no one in the company had any particular problem with people knowing their physical location by using their mobile phones. However, when I talked to other companies about this concept it unleashed fears of a 'Brave New World' where our every movement is monitored.

It is possible and totally acceptable to be as private in the Digital Workplace as you were in the Physical Workplace. It is your choice and an employer has no right to expect new deeper levels of disclosure in the Digital Workplace. The Nokia example is not an issue of privacy.

If you are at work, why shouldn't your colleagues know where you are physically if this helps them to work with you? You are still free to work where you wish but the details of your location can be shared with colleagues through GPS. What you say on a personal level about yourself on the employee directory is your call, but the fact that you are working at home or in a café near your home should not be treated as private.

There is a huge difference between needing to know where a colleague is at any given time while at work and wanting to know where they are going on holiday, who they live with or even where in their home they have set up their home work station – because none of those pieces of information support their working practice.

Some people tell their colleagues their life stories (every day it seems) and others like to keep themselves to themselves. This is grown-up working. Through being online, our lives and whereabouts are generally better known than previously and authorities can track us most of the time. Whether or not you approve or disapprove of this is your call but it is not part of the working relationship. In IBF, we share some social items on a micro-blog we run internally; some use this while others prefer not to; sometimes I do but generally I don't, as I prefer to keep my personal and professional lives at a healthy distance from each other.

It's also worth keeping in mind that global organizations can fall foul of local legal restrictions if they assume that cross-company personal details can be shared. Let the employee or freelancer decide for themselves and do not force disclosure unless it directly improves someone's work and productivity.

It is possible and totally acceptable to be as private in the Digital Workplace as you were in the Physical Workplace.

Risk and security

There are crucial risk and security issues that can be exposed in the Digital Workplace. These need to be addressed but, as with the technology selection, I don't aim to address this explicitly in this book. These are both highly technical areas that require specialized debate and insight – not my area – but they are no doubt topics that will be covered thoroughly elsewhere.

Case Study

You can't put the genie back in the bottle

Kate Lister is the co-founder of Telework Research Network,[33] a US-based independent research and advisory firm that specializes in making the management case for telework, workplace flexibility and alternative workplace strategies.

Lister believes that teleworking practices change the very nature of work. *For the most part we still work in an industrial revolution kind of way – we tend to go to one place and crowd together because that's what we're used to. But technology has broken the tether between 'work' and 'place'. It's changing our concept of 'work' from a noun to a verb – it's what you do, rather than where, when, or how you do it.*

She believes there is a series of clear benefits to be derived from teleworking. *Telework allows companies to hire the best and the brightest – regardless of where they live. Through the flexibility it offers, it reduces stress and work/life conflict. It empowers people to do their best, increases engagement and inspires loyalty. It saves money – lots of it – for both employers and employees. And it's good for the environment. As a result, the orgainzations that 'get it' are more competitive, more profitable, and more sustainable.*

While saving money has not traditionally driven telework initiatives, according to Lister, the recession has changed that. *There's more emphasis on telework right now because of the recession – organizations have been willing to give it a try because they were desperate for cost-saving solutions. But I often call telework the 'solution to the problem de jour.' Before the recession hit, talent and labor shortages led the list of management concerns – telework offered a solution. The environmental benefits of telework were another key driver before the recession because it offers one of the easiest, cheapest and most effective*

ways for companies to reduce their carbon footprint. Continuity of operations has traditionally driven organizations to telework too – it's the cornerstone of the US federal disaster preparedness plan.

So while economic issues have been a catalyst for telework in recent years, the pre-recessionary drivers will likely continue to move it forward. There's also the reality that once people have a taste of telework they don't want to go back to the cubicle farm – you can't put the genie back in the bottle. I think that's where we are right now.

In studying and researching many telework programmes, Lister has identified trust between managers and employees as the key criteria for success. For telework to work, managers have to stop managing by presence and start managing by results.

If you're going to have your people out of sight you have to trust them. You have to believe they're doing what they should be and the only way to do that is to manage by results. Management gurus have been telling us for decades that people want to be trusted, they want to do their best, and they want to feel a part of a greater whole. What they don't want is to be micromanaged. This is something that leading organizations understand. Why should you care if your top performer is playing golf at 4 o'clock on a Thursday, as long as they're meeting their goals?

With globalization and technology and mobility, people aren't at their desks anyway. When companies look around at their utilization of office space, it's not uncommon for it to be as low as 25 to 30%, because the people just aren't there. If you're a manager, you may not be seeing your employees anyway, so if you're not managing by results, then you really don't know what you're getting.

The fact is, the employees have already left the building. Space utilization of 35 to 45% is not unusual in a typical office building because the people are on the road, in meetings, at customer locations, and increasingly working at home or in what's become known as 'third places' – coffee shops, libraries, parks, etc. So results-based management is something that's needed regardless of whether people are working two floors, two miles or two continents away.

However, teleworking needs planning and support. This includes senior management buy-in and training at all levels of the organization. Lister explains: *First the executive team has to believe in teleworking and give it their unambiguous support. But even if it comes from the top, if middle managers don't embrace the concept, it's just not going to work. That's been proven over and over again.*

This isn't something you roll out and expect everyone to be on board. There are huge organizational and cultural barriers that need to be overcome for a telework program to be successful. On the organization end, at a minimum, the HR, IT, facilities, and finance people all need to be at the table from the onset. It has to be a coordinated approach.

The cultural barriers can be overcome with time and training. Managers need to learn how to manage by results, how to communicate, collaborate and maintain the 'esprit de corps' when their people aren't physically together. Employees need to learn what it's like to work remotely, how to use the tools, and how to stay visible, connected and motivated. This isn't rocket science but it isn't business as usual either.

The other thing is that I think companies have learned, and best practices have shown that, this isn't something you just do. It requires some planning. It certainly requires training at all levels; and again this has to come from the top down. Managers need to learn how to manage by results, how to inspire other people, how to communicate and collaborate. Employees need to learn what it's like to work remotely, how to use the tools and how to be self-motivated. They need to know how to communicate and to stay visible with both their bosses and their co-workers.

Top ten challenges in establishing a successful Digital Workplace

1. Fear at managerial level and among older and long-term staff.
2. Lack of physical contact between people.
3. Isolation that causes the loss of a sense of being part of the organization.
4. Lack of 'touch points' with manager and colleagues.
5. Low-grade technology that causes endemic frustration.
6. Command and control attempts in a new flatter structure.
7. Fragmentation and duplication because informal, in-person networks have been disrupted.
8. Two-tier workforce where some are in the Digital Workplace but others are not due to their roles.
9. Risk, security and insurance issues when work/life blurs.
10. Personnel issues due to extended hours of home working.

Chapter 15

Speed

What happens when everything is faster?

There is a well-honed advert from the world-famous Bourbon producer Jack Daniels that is all about taking life and work slowly and, as a result, delivering the highest quality results. Is that how they actually work at the Jack Daniels distillery? I have no idea, but the point being made to the 'always on–reply now' Bourbon target market is that there is an alternative style of working, an entirely different business culture.

When things become digital, time almost evaporates. We expect reactions from our technology – and increasingly from our colleagues too – that are close to instant. We panic when Internet connections run a little slow, when programs stutter, and when those we communicate with take some time to respond. Speed of working can now be almost immediate and we will impatiently swap from email to instant messenger to text to online app in order to gain that tiny sliver of extra speed.

Sherry Turkle, in her groundbreaking book *Alone Together* says that, just because we *can* work (and live) fast does not mean we *should* operate at this hyper-speed.[34] And Jaron Lanier in his manifesto *You Are Not a Gadget* challenges us to appreciate the limits of rapid response; this from a man at the heart of Silicon Valley.[35] The trouble is, when a potential new customer asks for information, a pause for quiet reflection does not tend to bring in new business.

We are all 'time travellers' now

In the Digital Workplace we can complete tasks at speed, receive reactions instantly and complete decision-making cycles within minutes, if we so wish. A manager I know received an email the day before Thanksgiving from another manager at a similar level who had decided without any justification that a clear reply from a small team was needed *that day*. The manager in question was on holiday and others prevaricated for a few hours causing the manager asking the question to sign off for Thanksgiving with an abusive, unconsidered email, no doubt partly down to work pressure – all because he had the concept in his mind that this email required an impossibly rapid response. A more considered reply after Thanksgiving simply wouldn't do. The new rules around what speed is acceptable are being redefined, often with negative consequences.

Meeting deadlines, reacting to urgent issues fast and resolving tiny problems that can escalate unless completed at once – these

Human beings need, as Jaron Lanier says, to remain the designers of our work and life. Otherwise, the machines we so love, will assume control without us even realizing.

> We panic when Internet connections run a little slower, when programs stutter and when those we communicate with take time to respond.

working habits remain highly valued because speed is helpful in each case. What has become blurred though is that a culture is emerging where an immediate response to everything becomes the 'new normal'. This not only creates an unnatural workflow but excludes the added value that comes from the nearly extinct habits of consideration, reflection and, in the end, thinking.

As Sherry Turkle also says, just because we have the technology, does not mean we have yet developed the maturity to use it rather than to allow it to use us. The artificial headline-grabbing challenges to technology from the likes of Thierry Breton, CEO of IT company Atos, who has decided to ban internal email, are futile.[36] We do not want a ban on the technology but instead we need to develop a culture that uses technology appropriately. Email is perfect in many situations, as is micro-blogging and instant messaging. What we need to reintroduce into the work equation is time. We need to take time where it is required, either because we are currently engaged in another task that needs full concentration, or because the communication will in fact benefit from pause, some discussion, possibly verbal or in person, and proper thinking.

What the Digital Workplace has produced is an option for speed to become normalized and, if we allow this new unhealthy and unproductive default approach to become standard (and also factor in the danger of 'addiction' covered elsewhere in this book), we have a cocktail that will produce severe health issues as well as deeply dysfunctional new working environments, both digital and physical.

The Digital Workplace flows to wherever we allow it. That is part of its power and capability but human beings need to remain the designers of their work and life. Otherwise, the machines we so love, will assume control without us even realizing.

We need to take time where it is required, either because we are currently engaged in another task that needs full concentration, or because the communication will in fact benefit from pause.

OUTSIDE
WORLD

Chapter 16

Customers, suppliers and external relationships

How has the Digital Workplace changed the landscape?

On a broader level, the online world is changing the way many markets and businesses, as well as non-profits, interact with their customers, supply chain and the worlds they serve. There are new types of conversation happening across every industry and these enable consumers to have a voice. Dialogue is opening up and what people think of you has become more and more visible, immediate and important. Your organization and services are accessible in ways that were formerly impossible. We can book a holiday in an instant and, as a part of the decision-making process, view real-life experiences from people who have already holidayed with the company we are considering going with, on a host of different sites (including the holiday provider's own site).

If your organization is poor in any aspect of its service, everyone will know about it again... and again... and again. Customers can follow you on Twitter, access your Facebook page and comment freely. When marketing guru Seth Godin bought

If your organization is poor in any aspect of its service, everyone will know about it again... and again... and again.

a new Apple Macbook, there was a disconnect between the promise from the store and his actual experience. Within a few hours, one million subscribers to his blog knew the whole story in detail. Ouch!

How has interaction changed?

The Digital Workplace is part of this continuum of open accountability and visibility. The managerial strengths and weaknesses of organizations tend to be exposed during shifts from the Physical Workplace to the Digital Workplace. Equally, staff have the ability to connect with the outside world of customers, markets and suppliers via the social network sites used by those people and extranets accessible only to specific groups. In this way, the boundaries between inside and outside begin to blur. Pacific Gas & Electric, a Californian utility, had received some bad media coverage in recent years and set up a new blogging channel where staff could

> A recent survey about virtual work sponsored by Regus reported that in **62.5% of large enterprises programs** involving 'new ways of working' had been rolled out, with only 8.5% stating there was no such program at all[37]

communicate easily with customers in an open conversation; this immediately became more popular as a destination for staff than the intranet.

How customers become advocates then employees

As mentioned before, in IBF we have had several customers who have joined us as freelancers, then become staffers, and some have even gone back again in the other direction. I often forget who has been a customer and who hasn't. Yes, we are a close-knit industry and know a lot of organizations and teams, but it is interesting to observe the connections as we move between conversations with staff, customers, and the wider world of non-customers, media consultants and students. This way of working enables staff to become advocates and subject matter experts more easily and they enjoy the experience. Are they 'trained' to represent IBF to the world outside? Not at all. There is no need; they seem to soak up the atmosphere and just 'know' what to say or not say. If something crops up that needs expert or leadership comment, then the person best suited is given a heads up and they will step in.

WH Smith: Ask Doris

Ever had the feeling you've heard that before? WH Smith had a very 'go to' person called 'Doris in HR' who could answer any question. So, each time a question came in via phone or email she would post her answer on the intranet. To ensure that this was personal and trusted she called it 'Ask Doris.' Every time the same question came in, she could point them to the intranet. Or if it was a new question, she would add it to the site. Job done!

Are relationships stronger or looser?

Relationships are generally stronger when an organization acts openly and consistently. If you use the new channels but communicate mixed messages, then you take a step backwards. But if you get this right, it seems to feed loyalty and engagement, and make people feel like part of your community – which they are in practice.

The opportunity is there to deepen ties and connections and to join conversations across the organizational divide. The threat is that people will talk anyway whether or not you choose to participate. Plus, you can do and say the wrong thing and will then suffer a bruising fast and in public. Nestlé got into hot water on its Facebook site because they tried to prevent people tampering playfully with their logo. The 'corporate identity police,' old hands at beating up people internally for any transgression to the sacred brand, found that on the outside this controlling hand could end up smacking you in the face, which is what happened. I read some of the Facebook backlash but got bored after six pages of negative comments!

> Relationships are generally stronger when an organization acts openly and consistently... if you get this right, it seems to feed loyalty and engagement, and make people feel like part of your community.

Employment, freelance, ad hoc project people – all much the same

The boundaries between staff, regular freelancers and ad hoc providers are starting to blur. In IBF, we have ad hoc providers, who do little paid work but who still contribute to our communities, and equally we have staff who say virtually nothing in those forums. Each way is fine. Everyone gets access to the same data, more or less, and has similar Digital Workplace experiences.

So do the staffers feel more a part of IBF? Yes, they do, because as a company we involve them more deeply in what we are about and where we are going. We choose to include them and actively to exclude freelancers at times. This is not to push the freelancers away but because the staff have a different relationship legally and in working terms with IBF. It is important to acknowledge that fact deliberately and to make the staff know that they are special. It's a subtle yet important point.

Chapter 17

Productivity

Faster, cheaper, better

Every organization starts gasping with excitement when talk turns to improving efficiency and productivity and driving down costs. CEOs, CFOs and everyone else with any clout carry in their heads the same organizational nirvana of their business or public sector body working better, more simply and more effectively.

This manifests itself in myriad ways. Take, for example, the image of huge trucks streaming across the outback of Australia, full to the brim with ice cream from Sara Lee, the global food manufacturer. Some shipments arrive late and incorrectly loaded, while others hit the mark every time. For Sara Lee, getting their product deliveries handled by the best, most reliable hauliers matters. Late ice cream doesn't taste that good.

At the regional offices and back at the headquarters in Holland, the daily facts around what gets delivered, to whom and when are all logged via intranet team sites utilizing simple-to-use collaboration tools. Everyone with an interest can see what is actually happening on the ground in Australia and these real-time data are used to optimize the Sara Lee business. This is just one example of the way in which digital working can enable multiple people both inside and external to the organization to access and use information in order to improve services.

This information is all part of the Digital Workplace for truck drivers, warehouse managers and supply chain managers both locally and centrally. Whether located in the office, at home, or at a pit stop en route, a swathe of staff and contractors is able to share data and facts, all thanks to a secure Internet connection.

When we think of working in the Digital Workplace, we think of collaboration, connection, communication, flexibility, and these aspects are all true. But the Digital Workplace touches work everywhere. Moving offline processes and data into digital workspaces allows them to be streamlined and simplified and the delivery cost of that process falls – sometimes to close to zero.

In British Airways, the penny dropped early on in the digital story. Staff are in the air, on the ground in terminals and hotels, and seldom (if ever) in the Physical Workplace. A centrally available portal was created a decade ago, so that flight crews could see their rosters, change shifts, book holidays, view payslips and so on. In the past, this had all required manual effort, face-to-face conversations and had been saturated with inefficiencies. This example of a Digital Workplace contains a range of digital processes and every transaction by a staff member is fast, cheap and, crucially, *better* than before.

Research conducted in 2010 in the UK suggested that **73% of managers** believed flexible working made their teams more productive, yet 52% of people who worked flexible patterns thought it had a 'detrimental impact on their career progression' [38]

Get rid of the boring stuff – focus on what matters

So much time is wasted in organizations on processes that don't matter. Why tell your boss in an email or even in person that you want to take a week's holiday next September when you could just log that request in a comprehensive system that contains everyone's holidays? Jams can be spotted, errors picked up quickly and the boredom of wasted time discussing it discarded. Or expenses management – why not move this online, simplify it and link it into the HR payment process with funds paid directly into staff accounts after a two-second sign-off by the authorizing manager (if needed)?

And then, when that's done, locate the process in the Cloud or some repository that is accessible from anywhere, and we can all rest easy and forget another aspect of modern work that is just mindless.

Plus, there is another major upside. The time released and, perhaps more importantly, the energy and focus enabled by essentially deleting a mundane repetitive process, may give that employee or contractor a clearer view into their work. I find that the more tedious distractions I experience, the less effective I am in my work. When we have an open, unrestricted perspective on our day, we become free to go where we need and produce what is possible; and we can thank someone on high that our employer has removed the shackles so that we can just get on with our 'real' jobs.

Being personally productive is exciting

I get a real buzz from driving my personal productivity through light, flexible technologies. On an average morning en route to the Physical Workplace, on the days when I am working there, a quick pitstop at Starbucks will prompt a rush of iPhone activity: Skype messages with several team members, a call with a customer, a chance to post some snippets of interesting links to the Yammer Member and IBF Live networks or to load some new Digital Workplace ideas to the Posterous life stream, which will automatically broadcast these as Tweets, plus a status update to Linkedin and a quick check of my emails, voicemails and texts – all for zero or close to zero cost. You feel like you've done a day's work in value terms before you even get to 'work'!

Likewise, one of my roles is to inspire/lead the new member activity across the business (sales really, although we try to dress it up in less brutal language). We have a pipeline of prospective new members that stands at around 1000 Fortune 500 or equivalent organizations globally. The team members across IBF are actively touching these potential new members daily and each moment helps the journey to membership along.

We log what happens in a shared, Cloud-based sales database from which we can see everything

Employees are evaluated on performance, not presence.

regarding that organization in one central place – accessible from a PC, laptop or mobile device from anywhere with an Internet connection. After a recent meeting with one prospective member who required benchmarking, within 20 minutes the data on what they needed was in our Cloud system; Angela in Ohio was already allocating the benchmarking resource, Lou in the Swiss Alps was starting her report writing, Dave in Canary Wharf was planning report delivery, while Fanny was processing the commercials from London and a bunch of others not directly involved were able to say 'Well done guys' via Yammer. By the time I'd finished my coffee near the prospect's offices, everything was in place; a week later the full benchmark report was delivered digitally to the (by now) new member. Normally, IBF benchmarking takes a little longer than this by the way!

This will save you bucket loads of cash

Driving productivity through the Digital Workplace makes sense for all the reasons above, plus it can also save you money in many areas. Manual, one-to-one, processes are pricey. With effective digital work practices, you need fewer offices and fewer people; this will drive down handling costs and can generate more sales.

I heard a wonderful story recently from a major financial services firm that was promoting a new smoother, simpler mortgage processing and loan application method for customers. This was all great until the application landed on a desk at the dull, old-fashioned office where the back office guys still toiled away; from that point on everything cranked back to the Dark Ages. So the company looked at re-engineering the process and training the staff. What a pointless activity! The best – and possibly only – way is to get rid of half or more of the staff, invest in Cloud technologies, encourage people to work mainly from home, or from smaller well-constructed offices, and to build a new way of working that actually introduces simplicity and productivity inside rather than just in the advertising copy that consumers see.

In IBF, when we identify a process or physical asset, we ask how we can either remove it or move it to the Cloud. The chances are that this will be faster, cheaper, better and more fun. So what is keeping organizations back? The truth is that driving productivity through the Digital Workplace is neither easy, painless nor risk-free. The BA story took vision, investment and training.

Few organizations as yet have the faith and confidence to really grasp how to work in the Digital Workplace but this is changing, particularly in countries with some government-level leadership, such as Holland and Finland. But don't wait too long, as your competitors are probably thinking as you are and these changes drive down the cost of delivering services and products. If they get ahead of you on this one they will grasp the competitive advantage.

Case Study

ROWE: Getting results with a 'Results-Only Work Environment'

The creation of a Results-Only Work Environment (ROWE) can potentially lead to greater productivity, reduced costs and a happier and more engaged workforce. Although there have been various management philosophies centred on results-based working, among the best known is ROWE itself, created by US-based HR mavericks Cali Ressler and Jody Thompson.

ROWE means that employees are 'evaluated on performance, not presence.'[39] They have total autonomy over their schedules and activities, which can mean as little or as much face-time in the office as required. According to Ressler and Thompson, technology is one of three components (along with trust and a flexible view of time) that is central to ROWE.[40]

Up to now the best-known example of ROWE has been at Best Buy, where Ressler and Thompson originated the idea. This has been written about extensively – both in Ressler and Thompson's own book *Why Work Sucks and How to Fix It*[41] and in Daniel H. Pink's *Drive*.[42]

Here is a quick summary:

In 2003, some departments at the HQ of US-based retailer Best Buy started to implement a ROWE program. Employees had complete autonomy over where and when they worked. They were not required to check into the office or even attend meetings. Instead, with expectations set by management and a culture of trust, performance would be judged solely on output. There was no concept of 'time' so ROWE was very different from,

say, a flexible working program. The evolution of the Digital Workplace was a central component of the program so that people could perform functions away from the firm's HQ.

For example, there were some manual processes (such as checking samples for fraud) which needed to be carried out on orders made on the bestbuy.com website. By implementing ROWE and the corresponding technology, which allowed these processes to be carried out off-site, metrics showed a rise in productivity and employee engagement within weeks. Perhaps most significantly, people who worked outside the office processed 18% orders more than those who worked inside, with an additional rise in metrics relating to quality.[43]

Despite some very stiff cultural opposition, senior management soon realized how powerful ROWE could be and backed the scheme. By 2008, Best Buy had extended ROWE to 4000 office-based employees. Ressler and Thompson claim the initiative has resulted in an average 35% growth in productivity and a colossal reduction in voluntary turnover rates, as high as 90% in some departments.[44] They claim ROWE results in an 'energized, disciplined, flexible and focused' workforce.[39]

Besides an implementation at Gap Outlet, most experiments with ROWE have taken place in smaller organizations. There have also been some fascinating experiments in American public-sector organizations, which are traditionally more associated with the culture of a 'fixed' working day and location.

In a ROWE program at the Human Services and Public Health Department at Hennepin County, Minnesota, a department of 2700 people have been transitioning into ROWE since 2009. Early results have been very favorable in creating better customer service through process improvement, as well

as better employee engagement and reduced costs. But it's clear that the cultural change takes time to embed.

Deb Truesdell, ROWE program manager explains: *One challenge has been letting go of our 'command-and-control' mindset. For years, people were rewarded and promoted for being able to manage adherence to a very rule-based environment. Also, a segment of our staff had never been given any real autonomy or allowed flexibility, and there have been some struggles around that. People did not necessarily have the confidence to believe in the concept and to actually be asked for the ideas about how to do the work differently. It all takes time…*

I have been surprised by resistance from some staff who appeared to be flexible and resilient. I have also been amazed at how well this works in many of our areas… One surprise is the number of 'converts' that we see: people who really didn't like ROWE but have completely changed their minds.[45]

Case Study

Microsoft Netherlands – New Work

'Het Nieuwe Werken' – 'The New Work' or 'New Way of Working' – has been a hot business topic in the Netherlands for a few years. Collectively, the country has some of the leading-edge thinking about working style, and the interaction between the Digital Workplace and the Physical Workplace. One of the prominent examples of a successful initiative has been Microsoft's Netherlands subsidiary, which was centered around a move to new offices in 2008.

The approach Microsoft took is interesting and a great example of how taking a holistic of people, place and technology can achieve successful results.

The New Way of Working program at Microsoft has resulted in an estimated:

- 30% reduction in building costs over five years;
- 50% reduction in telecommunications costs over five years;
- 21% reduction in carbon footprint since 2007;
- 40% improvement in employee satisfaction scores over five years.[46]

The roots were in a white paper Bill Gates had written in 2005 about 'Digital Work Style.' Gates found that: physical offices were not necessarily good for productivity; collaboration was key to getting closer to customers; and in order for companies to grow, they needed to provide opportunities for personal growth for their employees.[47] In Holland, a leading team of key stakeholders were selected around people, place, technology, and

> *"What they came to realize was that their own perceptions had to change in terms of how to manage, and in doing so required the use of output driven methods of measuring overall employee effectiveness."*
>
> Eric van Heck, Professor at RSM

sales and marketing. The latter was important as there was an explicit link between the initiative and the messages they wanted to give to customers about Microsoft products. The Rotterdam School of Management (RSM) was also involved in order to provide analysis on workplace trends and to measure various factors 'before and after' the office move.[48]

The major outcome of the program has been a move to new offices with no fixed desks at all and an accompanying range of supporting technologies, both devices and software, to enable location-free working either inside or outside the premises. There is also far more flexibility for employees to work when and where they want.[48]

Microsoft has been quite candid about what has worked and what has not. The change management program provided many challenges. The lead team recruited about 50 'change agents' within the business to 'spread the word' about the new way of working, but there was a disconnect between individuals' expectations and how the lead team saw their role. The change agents wanted more say in the detail of the program. Involving the firm's senior management was far more successful. They visibly gave up their individual desks and hot-desked around the business in a different part each day. They spent more time working from home so it became more culturally acceptable, and they opened up their calendars to all employees.[48]

Microsoft also acknowledges that it didn't necessarily get things right first time. For example, a new self-service hospitality tool was introduced which meant that staff members arranging meetings had to book their own meeting rooms, equipment, catering and even organize visitor car parking places. But there had not been enough training. Some employees tried to circumvent this by going to reception to organize things like this in the same way they were used to and this caused some tension. Eventually, a simpler online form-based approach was introduced.[49]

Other initiatives included an 'away day' for all employees. To register for this you had to fill out a short questionnaire about working style. From this, four types of employee were identified, which the team dubbed Entrepreneurs, Researchers, Idealists and Diplomats. They then derived additional separate communication strategies for each group to support the initiative – for example, they would send short instructional voicemails to Entrepreneurs but memory sticks full of documents to Researchers. Additional training on working style was also given – for example, sessions called 'Pimp my workstyle' and 'I love email and tasks' were provided for various groups.[48]

Meanwhile, RSM's analysis has proved interesting. This found that the initiative had increased mobility amongst the workforce, both internally and externally. They also reported increases in their measurements against four criteria: productivity, innovativeness, satisfaction and flexibility. The main cultural challenge they had identified was trust, particularly among managers.

Top ten benefits of the Digital Workplace

1. Huge reduction in building costs.
2. Ability to attract and retain best employees.
3. Healthier and more productive staff.
4. Large savings in carbon output and environmental impact.
5. Shift to an 'adult-to-adult' employment relationship through focus on outputs not inputs.

6. Potential for reduced headcount as positive changes take root across the organization.
7. Large savings financially and in terms of stress through less travel and commuting.
8. Ability to work effortlessly across regions in new virtual teams.
9. Platform of work that allows new technology breakthroughs to be adopted.
10. Closer and more flexible relationships with customers, marketplace and suppliers.

Close

DIGITAL WORLDS

Chapter 18

Digital disappointment

How can something this good feel so bad?

So, let's say that Google's Mountain View HQ is the ultimate experience of a Physical Workplace. Funky everything from the moment you set foot in there: before eventually settling on your chosen place to work for the day, you might jump on one of the bikes supplied for employees to get around the site, drop into one of the numerous eateries serving great food, wander through a huge aquarium area complete with loungers, stop for a game of pool. Are you at work or just having a good time somewhere fun and beautiful? Both, says Google. But if Mountain View is the pinnacle of 100 years or so of shaping and designing a working environment, how are we doing in the digital equivalent?

Imagine you are walking into the Digital Workplace. How does it feel?... *Too often, if this were a building, you would find yourself walking up a long flight of stairs... only to find it just stops, with no option but to leap across a huge gap to a new room, furnished and laid out in a totally different, but equally disconcerting, style... There are doors on the ceiling, walls that seem not straight... As you attempt to navigate this strange and uneven terrain, you are constantly halted by an odd crew of security guards, who repeatedly demand to know you who you are: 'What's your password? We'll need to see some ID.' You answer and move on, only to be apprehended by another. 'Can I just check, sir. Where were you born? Your mother's maiden name? What's your password – no, that is not correct.' All the while you are also being assailed by random messages in your ear and incessant requests for data. It's just an irritating chaos.*

Unfortunately, that is what the Digital Workplace feels like for many of us today. We tell everybody that technology allows us to do stuff anywhere, anytime and we are constantly being amazed by new functionality (remember the first time you could access emails on your phone?), but the truth is that most of the time the Digital Workplace experience is dreadful at the moment. If it were a building we would demolish it and start again! However, we can't do that in the Digital Workplace – it just needs to start to feeling and working better.

Does this 'digital door' open?

We should treat the Digital Workplace as we do physical spaces and hold a vision of an integrated, coherent world where one can flow effortlessly between one service or device and another.

Currently, the experience is awkward and uneven. All too often, you find yourself stepping from the sign-on page into an HR service that feels like a different universe, then attempting to enter another page... but at that point another password (which you've forgotten) is required, while all the time instant messages are popping up in another application... then you step into another room and lose connection totally. Organizations have not yet started to think properly about the design and architecture of the three-dimensional Digital Workplace world from the standpoint of either employee or contractor. What is required is effortless movement and ease so that Digital Workplaces can become truly productive, stress-free, secure and focused.

Take enterprise search. In most organizations, finding data or processes or colleagues is far too often a frustrating experience. It's like a physical world where you might spend 30 minutes hunting for the restroom, frantically searching for a sign and venturing up countless dead ends. We need the user experience experts and those involved with the emerging area of data visualization to start engaging with sculpting the worlds we are expected to work in digitally.

Think of the Digital Workplace as a three-dimensional digital equivalent of the Physical Workplace – but freed from the constraints of physical restrictions. In this new world, even time can be manipulated to some extent. Isn't this already starting to happen when someone adds to your shared Google Doc while you are working

Imagine you are walking into the Digital Workplace. If this were a building, you would find yourself walking up a long flight of stairs... only to find it just stops, with no option other than to leap across a huge gap to a new room, furnished and laid out in a totally different, but equally disconcerting, style... There are doors on the ceiling, walls that are not quite straight...

on something else? Or when a colleague connects with you through seeing what you are working on right now? Time is being flexed a little already.

Apple has always put user experience and user design at the heart of success in technology; hide the engineering and present something that feels so good you want it NOW and will queue for days to get it. Google has done this as well on a functional rather than an aesthetic level: it may not look great but, boy, is it solid and available!

Second Life was a brave first attempt at delivering a totally new vision of the Digital Workplace and, while it failed for a host of reasons (not the least of which was that it tried to just copy the Physical Workplace in crude ways online), it does show that the Digital Workplace we want is not two-dimensional (as most enterprise technology is now) but three-dimensional. Imagine a Digital Workplace that flows and bends at your will. We should have bold visions of what the future Digital Workplace will look like and then maybe the 'Digital Disappointment' can be overcome and the Digital Workplace will have us purring with 'Digital Delight.'

Chapter 19

Digital highs

Pinch yourself! It is this good – and getting better and better

In a café in London's Portobello Road in 1997, my then six-year-old daughter and I opened up my new Nokia 9000 – part of the hugely innovative 'Communicator' series – and accessed my company website. We were both struck dumb. It felt like a whole new world had opened up and, despite our different ages, we were each hugely attracted to what we were seeing and experiencing. This was not my first exposure to the Digital Workplace (that was when I first used an early computer in 1983 in a newsroom in Sydney) but it was a clear 'Digital High,' a moment when the power and reach of technology designed for work took my breath away. It was a sunny morning in a café I loved – and I was connected.

We are in the very early stages of the new digital world of work and there is much that we will come to regard as antiquated in the future, but what we can already achieve digitally is bewildering and should be a source of amazement. The reach of Wi-Fi and 3G is already extensive in the world and mobile phones with ever-richer services are stretching across even remote parts of Africa. We can already carry out

One mining company is designing a system that will operate driverless trucks in remote mining areas with the control of the truck happening from a technology system driven in another country; it can even be operated from the home of the 'remote driver.'

The physicality of work is decreasing and the digital world of work is only just beginning.

most work – aside from the areas I call 'untouched' in the next section – from locations of our choosing. Already the highs are intense.

But it's what will come next, and probably very soon, that is going to produce a steady stream of new 'digital highs.' One mining company I know of is designing a system that will operate driverless trucks in remote mining areas with the control of the truck happening from a technology system driven in another country; it can even be operated from the home of the 'remote driver.' When the first safely executed such operation takes place, those watching will know instantly that mining work will never be the same again.

When a factory manager based in Mexico wanders through a facility in India as a 'real as life' holographic projection, talking to those on the floor of the factory as if she were there in person, those experiencing it will feel that the future has arrived. When the video teleconference is just as tangible, and in some ways richer, than an actual physical meeting due to the quality of the three-dimensional rendering, we will just gasp at the quality and impact.

Much of the shift to the Digital Workplace so far has focused excessively on office space reduction as people begin to work from home. Boring! That is the least important effect of the Digital Workplace and is just one rather obvious and probably transient (as BT and IBM have found) stage on the journey. It is not what the Digital Workplace replaces physically but the new opportunities and ways of working that it is capable of creating that will count. The physicality of work is decreasing and the digital world of work is only just beginning.

A spaceship in our pockets

This book does not try to cover the technology road map for the Digital Workplace, which is changing by the day. How do we keep up to date? The shape will emerge from watching key players like IBM, Microsoft and Google, who are already engaged in fashioning the Digital Workplace of 2018. What they create for themselves will be the vision for their customers and this future is part of what leading technology firms are exploring now. It will involve being able to work anywhere, on any device, with effortless connection to others, using omnipresent high-quality audio and video. Language flexibility and time zone collapsing are also viewed as very important in global organizations.

It is early days on this journey into this new-style workplace. We are in awe of what we can do – but, at the same time, disappointed by what we can't. It's getting better fast but patience around this is a quality many organizations and people need to demonstrate.

The second half of this section collates some stories from the current Digital Workplace. Each case study has something to teach us in its own different way. Treat the next few pages as 'food for thought' when looking and planning your own Digital Workplace future.

Case Study

Upwardly mobile: How personal devices are changing the Digital Workplace

Key to the Digital Workplace are developments in smart phones. Even the most basic phones now feature, as standard, functions we could only dream of two years ago.

The explosion in mobile device use is starting to change the Digital Workplace within the enterprise. Many organizations are still at the point of thinking about how to integrate mobile devices, and large technology vendors are still tinkering with their offerings. Developments are led by what happens in the consumer space. With the introduction of the iPad and a number of rival tablet devices, the game is changing significantly. The introduction of iPads in the corporate workplace has caused a high level of adoption at senior level and therefore a huge rise in the understanding of mobile possibilities at senior levels.

A number of frontrunners are leading the way. Early adopters like Vodafone needed to get information to their staff in their retail outlets and, as Vodafone people naturally already had advanced phones, it was a low-cost and straightforward option to provide access to some of their intranet and collaboration tools.[50]

A more recent development is the growth of the internal enterprise 'apps store' which emulates the basic model of Apple's own app store for its iPhone and iPad. Research suggests that there will be exponential growth in

the number of apps produced over the next few years, both for consumers and in the workplace.

Research house IDC predicts that the number of apps downloaded globally will grow from 10.9 billion in 2010 to 76.9 billion in 2014, generating revenues of over US$35 billion. It's no wonder IDC analyst Scott Ellison predicts that: *Developers will 'appify' just about every interaction you can think of in your physical and digital worlds.*[51]

Given this backdrop, it comes as no surprise that IBM has developed its own enterprise mobile app store called 'Whirlwind.' Launched in the autumn of 2010, staff can choose from 400 different apps that they can add to their BlackBerry. These allow them to carry out core processes such as booking meeting rooms. The store also allows users to rate apps, and there is even an accompanying process for users to build their own applications.[52] Other companies that operate their own internal app stores include Google, Kraft Foods and SAP.[52]

Another emerging trend is the use of geo-location services within the enterprise. Using services like Foursquare and Gowalla, people check in at their location and this appears as a 'status update' on Facebook. Most corporate interaction with services like these tends to be for marketing purposes, but there are obvious uses within the Digital Workplace as well. For many years, people have updated their location within status updates, for example, on instant messaging platforms. This was often a reliable way of telling whether somebody was in the office or not. The automation of this process via a geo-location app on a mobile phone may be appealing, but there are also privacy concerns as mentioned earlier.

So far, penetration for this functionality in the enterprise has been limited, but 'location,' which helps track people and assets could be big business.

Schroders – I know where you are

Knowing where your colleagues are is useful as it enables you to say things like 'While you are at the Dorset office, ask John to show you the new drawings' or 'I see you are in New York at Customer X, do you have time for a coffee?' What Schroders did was simply ask people to update their status on where they were that day physically plus basic information on what they were up to. Short, simple and powerful.

There are already applications specifically to track people, assets and equipment, such as Ekahau. With this, anybody wearing a 'Wi-Fi' tag – essentially a small tracking device – can be located within a Wi-Fi network area. These are generally used in environments like factories and hospitals.[53]

Geo-location functionality also gives opportunities for using 'augmented reality' with mobile devices. Augmented reality (AR) is basically a mixing of the physical and virtual digital worlds, and has been defined as being 'real and virtual' and 'interactive in real time and registered in 3D.'[54]

An example of how this might work with the use of a mobile phone is being able to view the physical world through the device, with extra text-based information relevant to the location appearing on the screen. This has obvious opportunities for advertising and contextual information such as restaurant reviews, which appear when pointed at a particular building. However, so far experiments in the enterprise have been limited, even though software developers like Layar are already creating a '3D Augmented Office,' which at the moment is an

immersive tour through a virtual office, but has principles that could be applied to the Physical Workplace.[55]

Case Study

Micro-blogging

"We really like this idea of taking concepts that have worked in the consumer space and bringing them inside the enterprise."

David Sacks,
CEO of Yammer

Yammer, a US company that produces a micro-blogging platform for use in the enterprise, has enjoyed phenomenal growth since 2008, reaching around 1.5 million corporate users by the end of 2010.

David Sacks, CEO of Yammer,[56] explains:
As knowledge workers, we spend most of the workday online using email, creating documents or using other enterprise software. Yammer pulls this all together into one centralized place where everyone in the office can get together. It's described as a 'virtual water cooler.'

You've got people who are working in different offices, and they could be in different cities, even countries. There needs to be a way to coordinate all these people and the virtual office is a way to make them feel like they're all working together. The idea of a Digital Workplace is absolutely real.

Case Study

How Deloitte drives value from micro-blogging

After seeing how Deloitte's Australian arm had used the micro-blogging tool Yammer to facilitate internal collaboration, Richard Buck, the partner responsible for Deloitte Digital in the UK (covering IT and knowledge management) decided to initiate an experiment. Starting with only three registered users and no official roll-out program, Buck set up a Deloitte UK Yammer subscription.*

A first-time micro-blogger asked how he might help a client; he received 30 valuable responses within 24 hours, put these into a client presentation and won a new project.

The results were dramatic. Usage spread virally and rapidly. Six months later, more than 2000 (nearly 20% of the firm's UK partners and staff) were using Yammer to collaborate on business questions and issues. With over 9000 separate posts, and 160 specialist groups, covering everything from VAT to Olympic sponsorship, Buck is delighted with the results.

A typical story recounts how a first-time user from one of the firm's regional offices asked how he might help a client; he received 30 valuable responses within 24 hours, incorporated these into a presentation to the client and won a new project.

There are many challenges associated with using consumer technology in the workplace, for example, IT and risk concerns. One of the reasons the implementation at Deloitte has been so successful is because it was able to address these

concerns prior to the implementation, without the sometimes heavy-handed guidance and governance associated with online collaboration tools that can discourage use.

Before a critical mass of Yammer users had been reached, Buck had already squared the approach with the Head of Risk and also worked with Deloitte's IT department to ensure that appropriate security measures were in place.

One very important rule was that no client name or associated project data could be mentioned. A simple automated system to scan for appropriate words and an appropriate level of Yammer subscription (the best possible) were put in place.

A welcome message from Buck for new users on entering the system for the first time reminded users that they could not talk about clients. The result has been only one accidental mention of a client name, and this was picked up and removed almost immediately.

In many ways, this 'light touch' governance may prove to be the optimum approach to getting value out of social tools in the workplace. In fact, as usage continues to grow, there may be a need to integrate Yammer with other established channels, such as Deloitte's intranet and the firm's IT 'help desk' function.

**This case study is based on an interview with Richard Buck in 2011.*

The future shape of enterprise technology – according to Lee Bryant

Lee Bryant is the co-founder of Headshift,[57] part of the Dachis Group, and a leading consultancy in the social business space. Lee has spoken and written extensively about deploying smarter and simpler social tools in the business.

What people say they want and what they say they do is quite different from what they actually do day to day.

Bryant believes that users' needs are not being served by many current corporate IT departments: *Some people like to have control so they want to work in the way they want to work and they don't want to be told by a system or by IT that the perfect workflow is A to B, B to C, C to D, etc.*

When you try to find the perfect functionality or perfect workflow and then assume that, because it makes sense logically, people will use it, this is a fallacy. Observation shows that what people say they want and do is quite different to what they actually do day to day.

I've sat through meetings with senior IT people where they say, 'Look, we expect the users to hate us. We've never delivered anything that makes them love us and we don't feel that we're going to change that dynamic any time soon. So we deliver what we deliver, they do what they do and the world carries on.'

It's not regarded as a competence of IT to have an understanding of behavioral psychology or even just a basic common sense understanding of how the world works and how people work.

Bryant points out the divergence in approach between corporate IT departments and developers working in the consumer space.

These days all consumer technology starts at the opposite end. It all starts with behavior, experience and outcomes, and then treats the technology bits and pieces, the products and the functionalities almost as a means to an end to get there. If you look at Facebook, they have rolled out some of their biggest features with teams comprising around 15 people, and yet they run this absolutely enormous operation with huge amounts of data. What they focus on is just the user experience. That's what they do, that's their job and then they find technologies to back it up.

Internal IT functions start from the opposite end. They start from, 'Okay, we need a content management system. What's the most number of features we can think of that we should have in a content management system?'

Internal IT has no competitors and it has no predators because its competitor should be consumer technology on the outside. So if I can get my job done better with Facebook and Wikis and social networks, why the hell shouldn't I?

Consumer demand itself is also likely to change the role of IT departments.

There's been a build-up of pressure from users who continually say, 'Look, at home, on my Mac, I can not only use a Wiki, I can probably launch a Wiki from the command line. I have Google, I have Facebook, I have Linkedin. It's simple, it's free. Why do I have to put up with this appalling level of really expensive technology which totally disempowers me in the workplace?'

Bryant believes these pressures will contribute to how IT is organized.

In the next five years I think IT will see a split of functions and velocities. So I think you're going to have core, mission-critical system IT that will continue to be conservative and, indeed, should be, and will protect the kind of operational crown jewels. I think there are going to be the plumbers who just keep the networks flowing – with no more power in the organization than building facility management people today.

I think all the rest will really migrate from the basement, metaphorically, into the business and work alongside business managers as technologists or engineers, or whatever you want to call them, who deliver business applications and business value.

In the end we'll see underlying platforms managed by IT and apps managed by technologists in the business, and then what you'll see is somebody coming to work and choosing 'Well, I need this bit of tool app, I need this productivity app, I prefer this content management interface to that content management interface, and so on.'

Nearly there? The race to replicate the face-to-face experience

One direction that collaboration technologies have taken is to try and simulate the face-to-face meeting by creating equally immersive experiences in a virtual setting. This has spread in several different

technological directions. One example is holograms. For example, if you go to Manchester airport in the UK and visit Terminal 1, you can hear the current customs requirements from a pair of virtual holographic security guards.[58]

3D worlds

Another direction is the complete 3D alternative universe, similar to those that have emerged in online role-playing games like 'World of Warcraft.' There have been well-documented workplace experiments with Second Life, in which users interact with each other through simple 3D avatars, although these are mainly for leisure purposes. On the back of considerable media attention, in 2009, Linden Labs created a special enterprise product, Second Life Enterprise, which was designed to sit within an organization's firewall. Elements included virtual conference centres and a list of standard business avatars, plus there were plans to effectively develop an 'app store.'[59]

There is generally a perception that Second Life failed to live up to expectations in the workplace, and there is perhaps some truth in this. Its creators Linden Labs were forced to close their enterprise product in 2010.[60] However, 3D worlds have continued as a mainstream workplace offering, mainly via IBM's Lotus Sametime 3D suite of tools,[61] while other companies like Sun have developed their own experimental offerings.[62] Generally though these developments have not experienced the take-up originally predicted.

Telepresence

By far the most successful replication of the face-to-face experience has been the growth of telepresence. This technology has been around in one form or another since the early 1990s, but it only really took off when large corporates such as Cisco got involved in the next decade. People were also familiar with it through its appearance in television programs like *24*.

Today, telepresence works by trying to simulate a face-to-face office-based meeting as closely as possible. Two telepresence rooms in separate locations are connected so that participants face each other. High precision video and audio make it an immersive experience and, with cameras hidden at eye-level in the screens, it means that participants effectively look each other in the eyes.

The solution is often marketed as a visual collaboration aid that can reduce travel costs and reduce carbon footprint. Cisco themselves have rolled out more than 500 telepresence rooms internally, spread over 200 cities across 50 countries, and this has made a huge contribution to reducing the firm's carbon footprint.[63] Other companies like Arup have seen reductions in travel expenses of 45%.[64]

From **You are Not a Gadget** by Jaron Lanier

It's people who make the forum, not the software. There is huge room for improvement in digital technologies overall. I would love telepresence sessions with distant oudists [an instrument Lanier plays and loves] *but once you have the basics of a given technology in place, it's always important to step back and focus on the people for a while.*

Telepresence is now set to move increasingly into the consumer space. Several hotel chains have already signed deals and are rolling out telepresence rooms to many of their major locations.[64] Even more significantly, in late 2010, Cisco have started to target the home consumer market with an offering called ūmi telepresence, an experience which they claim will be 'transformational.'[65]

In many ways telepresence can be regarded as the winner in replicating the face-to-face experience. With the market now maturing, the technology is likely to get better and better and the experience even more immersive. So are we nearly there? No, but we are closer than we think.

The new world of work: how one Internet firm built an app for global talent

Some large web technology companies are now starting to use service marketplace sites for recruitment purposes. oDesk[21] is working with four of the largest 10 web companies in the world to tap into a global workforce to source staff in different geographies.

Gary Swart, CEO of oDesk, describes one of the applications they've built for a leading 'web company' that was looking for a solution to translate copy into different languages.

> We took a system that was routing the work and routed it directly to the end worker.

They were asking: 'Is there a way that we can tap into the global workforce for hard to find jobs in different geographies?' So, for example, they might need five people in Belgium that speak Dutch and Flemish, and 15 in Thailand and could we help?

We have, in fact, developed integrations that enable one client's system to talk to our platform, and we've built a flexible bench of pre-screened, rated and ranked, tested and trained talent. These contractors work as needed, on demand, directly through the interface that we've created with our client.

One client gave us parameters of what they were looking for; we invited 1000 contractors and, of the 1000, 800 applied; of that 800, 750 took the test

requested by the client; 500 passed and were able to start working right away.

What the client used to do with this specific work was to go to a big firm, who sold them tools and processes, who in turn would go to an agency, and that agency would bring in the end workers in each country. It would take weeks to identify who those people were and to get them contracted and then trained on the work.

We took the system that was routing the work and put the end worker in control. Now, a messaging platform says to the worker, 'Hey Steve, you have work assigned. Are you in or out?' And you can accept it right on your cell phone, or you say, 'No, I don't have time to do it, I'm at a rugby match.' But as soon as you accept it, you are now on the clock. Or you can pass it on and it goes to the next person in the queue.

The other benefit through this system is that all the work is reviewed by a quality assurance (QA) layer, and that QA layer assigns a score to each individual worker. This score becomes the quality rating for the work they deliver. Was it the right content? Did it have spelling errors? Did they produce good work and was it on time? This is a highly tuned work routing platform between the client and oDesk.

Case Study

Encouraging connections: how IBM (and others) are using social tools in the Digital Workplace

Of course, 'social tools' existed in the Digital Workplace well before the term 'social' came into such widespread use. A core feature of intranets has always been a searchable employee directory and, to a certain extent, some degree of informality has existed, for instance, encouraging hobbies and interests to be listed on individual staff profiles.

Inside IBM

Early adopters included IBM. Their 'Blue Pages' was originally conceived as a replacement for the firm's phone book but, as early as 2002, it was already storing information such as fields of expertise and project or team data. Individual 'Persona Pages' added contact data and pen portraits and the firm actively sought other useful information that would enable people to locate the right people across the global firm.

> The idea of the site is for IBM staff to have a rich connection to the people they work with on both a personal and a professional level.

But the world of social networking has had a significant influence on workplace technologies. Collaborative tools are now much more likely to be orientated towards connecting people together and displaying their behaviors. Internal tools

often now feature elements of Facebook, such as lists of relationships ('friends') to different colleagues.

IBM's response to the challenge of social networking has been to update the original Blue Pages and establish their platform SocialBlue, formerly known as 'IBM Beehive.' This has many Facebook-style features such as status updates, the ability to share photographs and set up events. In their own words, the idea of the site is for IBM staff to have a *rich connection to the people they work with on both a personal and a professional level.*

One brilliant little twist that IBM have added to the platform is the ability to add lists, known as 'hive fives.' These can be business-oriented, such as, five thoughts about a project. The format has been highly successful in getting IBM-ers to submit content.[66]

Sabre Town

Another innovator was global travel company Sabre. In 2008, they created an internal social networking platform called Sabre Town which featured individual pages for each employee, with a fresh look and feel that felt more like a consumer-led website than an internal company resource. Employees can add elements such as photos and blogs as well as answer light-hearted questionnaires on their profile.[67]

The most powerful element is a question and answer functionality that allows users to submit questions to the entire organization. An algorithm based on individual profile information and previously answered questions automatically sends the question to those whom it considers to be the 15 most 'relevant' people to answer that question. The results are powerful, with 60% of questions being answered within an hour.[68]

For answering questions correctly, and if the answer is rated highly by other users, individuals are awarded 'karma points,' which then translate into the ability to unlock extra functionality such as adding more pictures to your profile.[69]

Sabre: Karma points

How do you get people to share what they know? An existential puzzle of modern organizations. Turn it into a mild competition as the guys at Sabre Holdings have done. Those who share and have their knowledge used, gain 'karma points' and people compete for the status in league tables. Silly, simple and effective!

Another organization that has devised some unusual and interesting twists on social networking is US online retailer Zappos. It has a unique company culture that places a high value on employee engagement and subsequently has very low staff turnover. There aren't many large organizations that choose to employ a Chief Happiness Officer or that have a third of the company actively tweeting.[70] One little trick they have employed is 'The Face Game.' Every time an employee logs on to the Zappos intranet they are presented with a random photo of a fellow co-worker and five names. They have to select the right name to match to the photo and, although there are no penalties if they are wrong, the scores are logged and noted. The individual is then automatically taken to the individual's intranet profile to encourage connections between staff. Of course, in theory, this should make the Chief Happiness Officer's job that much easier![71]

Case Study

Transparency in the Digital Workplace: innovations at Love Machine

One of the most interesting innovators in the Digital Workspace area is Philip Rosedale, founder and CEO of Linden Labs, the creators of Second Life. Rosedale's latest venture is Californian-based start-up, Love Machine Inc. The company is not only innovating in the Digital Workplace through products but also in the way they do things.

Some of Rosedale's ideas seem to stem from his belief that 'there is always a win with transparency.' For example, when he was at Linden Labs he used to send a quarterly anonymous electronic 'CEO survey' to all his staff, which consisted of three questions:

1. Do you want to keep me or get a new CEO? <yes,no>

2. Am I getting better at my job? <yes,no>

3. Why?

A day later he would send everybody a link to a graph of the results for the first two answers, while taking on board any comments in the third.

In the early days of Love Machine Inc, they also built a special 'Bullshit Meter' button for virtual meetings, which anybody could operate if they thought somebody wasn't being entirely truthful.[72]

Messages of love

Love Machine's main product is also about being transparent. It is a hosted online employee recognition system called 'The Love Machine for Work,' which grew out of a system that is used internally at Linden Labs. Employees send each other 'love,' short positive thank you messages for doing great work or for helping each other out. Everyone in the company can see these and can also view a graph which indicates who has received the most positive feedback. At the end of a certain period employees get a proportional share of a predefined sum in gift card vouchers based on the amount of thank you messages they have received. As the system is online, it is perfectly suited for virtual companies and is already being installed in some large companies, such as Yelp.[72,73]

Employees send each other 'love,' short thank you messages for great work. Everybody can see these on a graph which shows who's got the most positive feedback.

Open design process

What is perhaps even more fascinating is the method by which the firm has developed its products. The founders have tried to fuse some of the aspects of Open Source projects with those of a freelance marketplace. They have taken their entire software needs and chunked them up into much smaller projects, putting these into a workplace environment that is open for anybody to see online. People from all over the world then bid for this work, receiving an SMS message to say their bid has been accepted.

All in all, the company has spent around US$360,000 on work spread around 100 people located in many different countries. All the code is open for everybody to see, which allows developers to keep up to speed. There is also a common development 'sandbox' which has a powerful chat facility (the 'workroom,' which is also available to view to the outside world), so developers can seek help from others if needed. According to Rosedale, this helped because: *the feeling of working together on a project in this way must be like being in the same room together.*[72,74]

The advantages to the company have been what Rosedale refers to 'an absence of ego issues' and this has allowed experimentation. He also says the quality of the programming code has been good enough for the company's needs. Moreover, developers have been able to set their own pay rates which works surprisingly well. Rosedale comments: *A critical discovery was how well and reasonably people behave and set prices when in a supportive environment with a high degree of transparency.*[72,74]

The company also takes an interesting view of its Physical Workplace. They have opened 'Workclub,' which is not only their functioning office, but also acts as a café. Other developers are allowed to drop in and work there, whether they are working for Love Machine or not, and they are experimenting with the idea of allowing students to use the facility in return for helping out here and there.[75]

Top ten tools for a successful Digital Workplace

1. Ability to access all work tools and systems from anywhere.
2. High-grade, advanced intranet, including all HR systems.
3. Fast access to all colleagues, with rich data on who they are and where they are right now.
4. Reliable technology devices that enable fast Internet connectivity – mobile, tablets and laptops.
5. Enterprise micro-blogging services with related connection tools.
6. Collaboration services with clear future road maps for easier upgrading.
7. Quick-fire teleconference, video and web-conferencing.
8. Access to all Customer Relationship Management and related database connectors.
9. Ability to work offline and the freedom to 'switch off' as needed.
10. Multi-language options in companies with that need.

OUR WORLD

Chapter 20

Travel, environment and demographics

Globalization

Technology, like health, is an area that has no geographical boundaries. If I can connect with you through instant messenger at the next desk, I can do this just as easily with anyone, anywhere in the world, aside from any time zone differences. The Digital Workplace has no physical limitations and therefore enables not only the globalization of work but also less distant changes whereby working in virtual teams becomes easier within one country. This provides options – and organizations like options.

> The Digital Workplace provides work options – and organizations like options.

Although this can be exaggerated, the global talent pool also becomes more available. Moving work around becomes far easier since staff need not necessarily be relocated with all the associated costs of that. Also, when people do move physically they can effortlessly stay connected to those they previously worked with in their former location. Outsourcing and offshoring of work has enabled workshifting to lower-cost regions, with all the positives and negatives of that trend.

Working from home is not 'a day off'

What I had not realized until I started researching the impact of the Digital Workplace and the trend that is known as 'workshifting' in the US, is how profound and far-reaching the benefits of this transformation can be to organizations, governments and society in general.

A reduction in commuting leads to: far fewer days off sick; fewer road accidents and fatalities (most accidents happen during the morning commute); less pressure on the health service; higher productivity; reduced staff turnover – and the list goes on. It's extraordinary, which is why governments are getting passionate about this subject.

A 2007 report by the Consumer Electronics Association in the US estimated that telecommuting alone in the US reduced the country's carbon emissions by **14 million tonnes** a year[76]

That is aside from the whole glorious bonus of avoiding that awful daily commute. No doubt you'll have your own favourite horror story – of being stuck on a crowded train with no information and somebody's newspaper stuck up your nose, or an epic journey of biblical proportions because of a light dusting of snow on the tracks.

Thinking about it now, I find it difficult to conceive how we have ended up with an activity that is both such a colossal waste of time and a systemic way of de-energizing the workforce before they've even got to work, not to mention a drain on the world's fuel resources.

What a waste of money!

It has been estimated that, in the UK alone, 'work time' worth £339 million a day is spent commuting back and forth.[77] I'm not going to attempt the maths, but that's going to be an overwhelming sum totalled up for the year, and a completely mind-boggling one for the globe. Sometimes I wonder if large organizations are actually addicted to commuting hell.

Getting yourself into the office is such an ordeal that it is often regarded as a sign of loyalty. I'm thinking of those emails from company leaders thanking the people who struggled to get into the office on a bad weather day when, in fact, the sensible ones stayed at home and actually got far more work done. It's all part of the macho culture that views working from home as 'work avoidance.'

The next time there is a London tube strike (and there do tend to be quite a lot), instead of channelling anger at the tube workers, perhaps we should secretly thank them for allowing us to show big firms that it's perfectly okay to work via the Digital Workplace. In the same way, one of the perhaps surprising benefits of the 'swine flu' scare was that at least it gave some impetus to IT departments to ensure that their Digital Workplace infrastructure was in place just in case the worst came to the worst.

Thankfully, the Digital Workplace is starting to make some positive changes. In the UK, there is evidence to suggest that the rise in home and

A reduction in commuting leads to far fewer days off sick; fewer road accidents and fatalities; less pressure on the health service; higher productivity; reduced staff turnover – and the list goes on.

In 2009 Gartner predicted that telepresence would replace **2.1 million** airline seats by 2012, a sum which would cost the travel industry approximately **US$3.5 billion** in revenue each year[78]

flexible working is starting to cut into the average commute time. The UK Trade Union Council calculated that, for 2008, the average journey to work had fallen to a 10-year low of 47 minutes and 48 seconds. This is despite the figures for this rising every year from 1998 to 2006 to a record high of 52 minutes.[77]

In short, journey times fell more between 2006 and 2008 than the total rise in the previous eight years. In the same period, there were nearly 300,000 more people working solely from home; flexible working became more acceptable in blue chip companies; and we also saw the mass adoption of broadband.

All up in the air

The Digital Workplace is also reducing the need for air travel. It's a mouth-watering situation for CFOs who not only have an opportunity to slash overheads, but also in a way that is morally justified and popular. Reducing the carbon footprint resonates positively with leadership functions, employees, shareholders and the public alike. In fact, the only losers in this quadruple-win situation are the travel industry.

In 2009, Gartner predicted that telepresence technology alone would replace 2.1 million airline seats a year by 2012, costing the global travel industry US$3.5 billion.[78] Okay, you will always need some international air travel – so there's a long way to go before the executive culture of draining the hotel mini-bar and checking their video selection doesn't show up on their bill disappears entirely – but the figures are significant.

For example, in their 2010 Corporate Social Responsibility Report, Cisco makes an explicit link between using collaboration solutions internally, principally WebEx and TelePresence, to host a staggering 19.3 million hours of virtual meetings, an annual saving of 47,000 tonnes of carbon emissions a year and a general reduction of 12% of greenhouse gases since 2007.[63]

> Cisco held 19.3 million hours of virtual meetings in one year with an annual saving of 47,000 tonnes of carbon emissions – a general reduction of 12% of greenhouse gases.[63]

Chapter 21

Untouched?

What will remain unaffected by the Digital Workplace?

Over dinner the other night I kicked off the conversation with boring predictability by announcing that 'offices are essentially dead.' They are tedious, wasteful and past their sell-by date, I proposed. The immediate pushback from two managers in corporate communications was that they would love not to go into the office but 'that's where things happen.' We debated this for a while with me suggesting that maybe it's exactly the opposite: 'The office is where nothing important happens.' Anyway, love it or distrust it, the Digital Workplace is not and never will be the universal answer to every industry and some types of work will remain either wholly or mainly untouched.

Theatres, art galleries, creative agencies, clothes shops, hotels, restaurants, travel hubs, dry cleaners and holidays – the list goes on; some services will always require a physical presence. Others require a degree of physical connection with colleagues in order to work well – such as advertising agencies. And then there are those that involve producing something physical and therefore, by definition, require some form of Physical Workplace.

What is interesting is that while I can't see how a dry cleaners can ever become digital rather than a physical place where you drop your suit off, other kinds of business, like PR agencies, might well end up with partial Digital Workplaces. Some of their work can be done better away from the distraction of colleagues, in a space where it is easier to focus and talk uninterrupted; on the other hand, the Digital Workplace can operate just as effectively whether you are in the office or out of it.

In fact, while there are untouched areas of work, the changes that may result from the Digital Workplace are unpredictable. In my own company we used to run a face-to-face annual conference, which was quite successful, but we got bored with it. We replaced it with a 24-hour annual online gathering called IBF 24. How would a virtual conference compete with the power of meeting people in the flesh? The virtual version won hands down: thousands rather than 150 attended, from 50 countries versus three, from hundreds of organizations. The volume of content was huge online compared with that possible in person and the time people spent actively engaged ran to more than 12 hours on average versus a fraction of that when in person.

Theatres, art galleries, creative agencies, clothing shops, hotels, restaurants, travel hubs, dry cleaners, holidays, conferences – all places that are little touched by the Digital Workplace.

Touched by technology: Remote surgeons, digital co-workers and robotic personal shoppers

Most of this book has been concerned with the emerging Digital Workplace in the office, but there are many other kinds of working day being transformed by technology. Automation and robotics are infiltrating different sectors, albeit at various different rates and stages of advance.

Up until now the potential for change has tended to outpace actual adoption. A combination of cultural barriers, cost-effectiveness and limitations of the technology itself has often slowed the take-up of these cutting-edge technologies.

Here are some examples of non-office environments that have already been touched by technology, offering some interesting variations on the concept of the Digital Workplace.

Hospitals

Some of the most truly revolutionary Digital Workplace technology is happening in the health sector. Remote surgery, where a surgeon is not physically located in the same place as the patient, although still not that common, has been successfully performed many times. The surgeon's movements can be transmitted via fibre optics while the patient is operated on by robotics. Visual data from an endoscopic camera are transmitted back to the surgeon. The most prominent early example of this was the 2001 transatlantic 'Lindbergh Operation,' during which a gallbladder was removed from a patient based in Strasbourg by a surgeon located in New York.[79,80]

Meanwhile, there are moveable robot 'stations' such as the InTouch Health RP-7 robot, which allows a doctor to be 'remotely present' for diagnosis. Videoconferencing allows the doctor and patient to interact while devices like stethoscopes and ultrasound machines can be plugged into the robot to transmit data back to the doctor.[81,82]

The 'Tug' is a robot which is used for a number of simple tasks in hospitals, such as delivering medications and linen between departments, saving time so that trained staff can concentrate on patient care. The robot is sophisticated enough to be able to operate lifts and has voice capability so it can announce its own arrival. It is claimed that it has led to time savings and a more efficient service.[83,84]

Restaurants

Restaurants would on the face of it seem to be exempt from the Digital Workplace but, even in this field, technology is changing the way in which people work and interact with their customers. For instance, the fast-food industry uses 'Hyperactive Bob,' a robot and real-time data processing system

A combination of cultural barriers, cost-effectiveness and limitations of the technology itself has often slowed the take-up of these cutting-edge technologies.

that helps burger restaurants and similar to predict the food they need to prepare in order to meet demand.

'Bob' acts like an automated kitchen manager, interacting with the kitchen staff via a digital touch screen, keeping them informed of how much of each food item to prepare, based on information it is constantly analyzing. These data take into account historical consumer trends, how much of each kind of food is ready and how many cars are coming into the car park (information captured from a camera positioned on the roof of the restaurant), even estimating the likely demand for food based on the types of car turning up! This system has been trialed and is already in use in a number of outlets, including globally recognized fast-food brands.[85,86]

Meanwhile, waiters no longer take the orders at the Inamo restaurants in London. These are equipped with touch-screen tabletops with browsable menus. Although primarily a gimmick, customers can call up pictures of the food and just order whenever they're ready. It's also possible to order a taxi, get travel information for the journey home and select an electronic tablecloth design to suit your own taste.[87]

Factories and warehouses

Some of the automation equipment and processes in factories – for example, in the car manufacturing process – are incredibly complex. Sometimes your co-worker, or even your manager, might be a robot.

Robot company Head There, have developed Giraff, a mobile video-conferencing facility that can be remotely controlled via a laptop. The Giraff can 'see' and 'hear' both close and distant objects, move around the factory floor and has a video display so people can interact with the remote user. It has been suggested as a way for managers to be partly 'present' in non-office workplaces.[88]

Meanwhile, companies like Gap and Crate & Barrel are using complex systems of robots at their warehouses to deliver goods to customers, reducing the need for manual workers. For example, robots deliver the right mobile shelf to a human being, shine a laser beam at what they think the right product is, and then a human 'picker' selects it from the shelf.[89]

Retail

Online shopping has already transformed the retail sector but, in 2008, Japanese robot manufacturer TMSUK went one step further. They conducted a successful experiment where a robot, remote-controlled through a mobile device, was able to enter a Japanese department store and buy a hat.[90]

The housebound grandmother controlling the movements of the robot could view what it was doing through a video connection to her phone. Similarly, the company has also co-developed the T-34 robot, a security robot with a video link which can be controlled by mobile phone, which can even apprehend criminals by firing a net over the intruder effectively immobilising them.[91]

Top ten myths of the Digital Workplace

1. Working away from the office is a 'day off' – staff actually work harder in the Digital Workplace.
2. The office is an efficient place to do a day's work – it's not and never has been.
3. 'Water coolers' drive innovation – this has virtually never been the case.
4. The Digital Workplace means the death of working together physically – no, you just meet physically less often.
5. You need to meet in person to build trust – often the reverse is true.

6. Video conferencing is the answer – most people prefer 'voice only' contact as it is more 'true to life.'

7. The Digital Workplace only operates outside the office – no, we are in the Digital Workplace wherever we are working.

8. The block to success is resistance to change among staff; it's actually the middle managers who are most fearful.

9. If you don't watch people working, how can you manage them? A focus on results is much easier to manage.

10. The Digital Workplace is already quite advanced... it isn't, it's barely started.

Close

FUTURE

WORLD

Chapter 22

Different planets

We are still struggling to understand the relationship between the Physical and the Digital Workplaces. Being human, we must always be somewhere physically. So, when we are working, we are in fact always in a Physical Workplace. But, increasingly, that Physical Workplace may not necessarily be a space created specifically for work and it can flex depending on what we are doing that day. The bus, taxi, café, airport, lobby, hotel, home, garden, and so on, are not places designed to be workspaces, but we may nevertheless find ourselves working in them. However, these days, we will almost always be occupying the Digital Workplace at the same time, aside from when we are not connected – no phone, no portable device, just a notepad perhaps – but how often does that happen? Seldom.

So, the Physical Workplace and the Digital Workplace are almost always co-present and interacting. It is the flexibility and ways in which they can be customized to suit each part of the day that is changing. This is not an either/or. We cannot choose to work in the Digital Workplace and therefore decide not to be in some form of Physical Workplace; it is rather that the Digital Workplace is consistent and persistent, while the Physical Workplace is now the major variable. In the past, the Physical Workplace was the stable part.

The Physical Workplace and Digital Workplace are always present and interacting.

The analogy I use is constellations. In a galaxy – our galaxy, for example – planets orbit in certain patterns relative to each other. There are a defined number. These are the various physical options available and you can move from one to the other. But between the planets and connecting the galaxy is space: constant, present and holding all the orbits in motion. This is the Digital Workplace. It is everywhere and consistent and far more fluid and flexible than the Physical Workplace. We are human and we are *physical* beings. We are also starting an evolutionary journey into virtual worlds and ways of being. It is early days. We will learn and get better at using the virtual and we will develop new characteristics suited to a new way of being.

We will see the complete physical fragmentation of work as companies and businesses function without any physical centres.

2040: a glimpse into possible futures

If I'm fortunate, in 2040 I'll be 82. If I stick with the yoga and tennis and fairly careful diet, and let's hope the brain can still work well, then what will I see around me in the world of work?

Looking at future points is always tricky. Why 2040? Well, it is just over 30 years since I joined the *Newcastle Evening Chronicle* with a typewriter in a smoke-filled newsroom and now here I am logged onto the Net in Starbucks, plugged into a data set beyond imagining. So let's roll forward another 30 years and see what the step change is.

Fragmentation and integration

Work will be fully released from the physical constraints of locality in all aspects apart from those areas of work and life covered in the 'Untouched' section of this book. We will still have restaurants, shops, arts venues, hairdressers and probably manufacturing plants (although likely run remotely in the main). Whatever can be digital, will be digital.

The Physical Workplace has been fragmenting relentlessly over the course of the past decade and this process will continue. We will work from anywhere, anytime and this will involve an economic reshaping as new industries and services emerge. The collective entity of the corporation will comprise staff, contractors and supply-chain linkages, which can carry out work effortlessly and continuously, irrespective of locality.

This will result in the complete physical fragmentation of work as companies and businesses evolve to function mainly without physical centres. But, just as digital spaces have removed physical connections, so they will enable new connections and relationships. In one sense people working together will be increasingly separate physically, while on the other hand they will be brought much closer digitally. Live connections through voice and sight will be so rich that images of the people we connect and communicate with will be so vivid as to be almost

When you enter the intranet, or whatever follows it, you will sense the same buzz as when walking into an impressive office of a new employer today.

identical to actually sitting together. Emotion, touch, sense and smell will all be part of what we currently call video conferencing. But it will not be the same as being together physically. Timothy Leary was right: in the future, physical meetings will be sacred, rare, special. This depth of digital connection will create new 'virtual offices' where some people will meet to work. We will both fragment and integrate in equal measure.

Physical offices will become extinct, reshaped, fragmented

Businesses that are 'touched' (i.e. 95% of all organizations) will have dramatic changes to the offices they own, depending on their sector and region and culture. All will have much reduced real estate and HQs will shrink in size and exist more for 'cosmetic branding' purposes. Companies will drive towards cost minimization and this will result in steady reductions in office capacity. Oil rigs, drug manufacturing and automotive plants, and so on, will continue to require a Physical Workplace but, once again, these will be smaller and fewer in number. Some companies will share co-working spaces with other companies and new types of working environments will be created that offer viable alternatives to the concept of home working. Your first day at work may be entirely digital but will feel far more visceral than is possible today. When you enter the intranet, or whatever follows it, you will sense the same buzz as when walking into the impressive office of a new employer today. The digital and physical will work better together but what is clear is that the 'digital real estate' will increase and the physical will steadily decrease.

Deeper, closer, better human and family relationships

I asked my own now fairly grown-up daughters whether work had ever got in the way of my relationships with them. They both said that my work had never caused a problem; I have been around a lot for them and always physically present in their daily lives. They did comment that tennis had got in the way at times... but that's another story. If we can work when we want, where we want, then we can live wherever we like and the entire demographic of modern advanced societies, including the developing new economies, changes. The city versus country split dilutes and the general population will spread as areas of regional neglect become new economically viable areas again.

In the last two years I have spent parts of each week in the English countryside, in the Cotswolds. This is not at all as I expected. It's not really the countryside, nor the city, but some hybrid where you can experience living and working in areas of natural beauty. There is a growing capability

> If we can work when we want, where we want, then we can live wherever we like and the entire demographic of modern advanced societies changes.

for people living there to generate income digitally (from outside the local area) and this is enabling areas like the Cotswolds to add a new stream to their development economically. This is a new template because economic power is separated from locality. You can earn in a global market but spend locally. The implications of this for demographic planners and politicians are profound because populations can become distributed far more generally. The tensions of homes versus conservation will continue but the population maps of major economies will look quite different in 2040.

The other consequence alluded to by my daughters is that working families will be able to enjoy deeper, closer, better relationships. I recently overheard on a train a retired American corporate guy talking to his wife and friend about how he had hardly seen his children for more than a few hours a week while they were growing up, aside from 10 days holiday a year. As he said, that's a loss you can't put a price on. Flip that coin over and enable that same dad to see his children every day and enter their lives in a richer, much closer way. What is the price or value we can place on that? What would be the impact of that over time on the health of our society? My own parental experience is one of connection and closeness, which has helped build confidence, power and personality in my girls I am certain, enabling them hopefully in time to pass that character of parenting on to their own children.

Top ten future impacts of the Digital Workplace

1. Large office buildings will disappear and become relics of a bygone age.
2. Big global organizations will have tiny but lavish headquarters for symbolic reasons only.
3. Most people will work from a range of new facilities such as co-working hubs, flexible bases close to where they live or in company-owned hybrid centres.
4. The countryside in major economies will be re-populated, creating communities that are neither 'city' nor 'countryside.'
5. The blurring of work/life, along with 'connection addiction,' will become serious health issues akin to obesity in levels of attention and concern.
6. Family and personal relationships will be enriched by 'workshifting,' enabling parents and children to enjoy more actual time together.
7. Organizations will create new 'part-paid roles' for people well beyond the traditional retirement age.
8. Security, risk and compliance will become more complex and challenging for organizations as technology becomes 'always there, always on.'
9. Physical meetings will become special, rare and highly valued.
10. Governments everywhere will drive policy in this area due to the scale of the gains to be had from the Digital Workplace.

Chapter 23

Reflections

Maybe you are one of those people who reads the end of a book first – yes they do exist – and I'm sometimes one of them! So here are some sound bites… plus a few closing comments.

As I said earlier, this book says:

- Most work in most organizations is awful – and we all know it.
- The Digital Workplace offers an inevitable change that transforms work.
- Most leaders and companies are blindly wandering into the future.
- That you should understand the Digital Workplace, design it into your future and be among the winners.

What the Digital Workplace offers is something employees and contractors long for – influence and power over how, when and where they work. It does not necessarily make work easier but it does change its location and design, fundamentally altering our relationship with work. All the evidence shows quite comprehensively that the benefits for organizations and people are extensive, including improved productivity, morale, cost, retention, environmental effects, communication, health, absenteeism, leadership, and so on. This is a movement in the workplace that ticks virtually every box valued by organizations, individuals and society alike.

There are some dangers and downsides though. People will tend to blur work and life and end up working longer hours rather than fewer. Managers are struggling to manage in this more results-orientated environment and are unsure what their role is when they see people less and less in the flesh. There are risks from a security and data stance as more and more work moves outside the firewall. Questions arise over how much real estate is needed going forward and what the implications are of people using their own technology for work.

For me, the Digital Workplace has been a long time coming. Work and location are being separated and after my first day at the *Newcastle Evening Chronicle*, that is something I longed for. It turns out now that many others have also felt the same.

If used well, the Digital Workplace offers freedom and accountability. How precious is that! We have embraced this shift in my own company and will never look back. We tackle the challenges it presents – such as the need to 'meet' when there is no fixed office – but we invariably find these issues easy to navigate. Over centuries, many of the major changes in society have centered around where work takes place. From agricultural work to industrialization and now to knowledge working – from the countryside to the city and to the 'pretty much anywhere you want' times we live in today.

Is the Digital Workplace some sort of 'workplace spiritual nirvana'?

My own belief is that humans are on a progressive evolutionary curve. Over centuries we grow and evolve as a species. We face compelling and possibly insurmountable challenges as humans, but if we track back over 10,000 years we can observe that we have grown inexorably in capability and knowledge. I tend to see the positive in virtually all situations, a trait which can seem either optimistic or just plain naive, but that is my character.

When this sea change in work begins to move through our societies, we will find more opportunity to move upwards within the 'pyramid of needs' as defined by the psychologist Abraham Maslow in his 1943 paper *A Theory of Human Motivation*.[92] Once survival and physical needs have been met, and the requirements of community, relationship and work are in place, then we can start to look at uncovering a deeper sense of ourselves as human beings. We can begin to find time for the two questions that have preoccupied me since I was about nine years old: 'Who am I?' and 'Why am I here?' The latter, in particular, is a question probably without final answer but just the opportunity to ask it with time and space is a true gift.

Am I saying that the Digital Workplace is some sort of 'workplace spiritual nirvana'? No, it isn't an elixir of life but it does possess the capacity to restructure people's lives, generally in a positive way, and this shift enables us to release and discover some internal and external freedoms that seem to support us as human beings. This is, in many ways, an odd and unexpected but hugely encouraging by-product of the Digital Workplace.

I am finishing writing this on a train to Durham in the north of England, with my youngest daughter Rose beside me; we are travelling to see a university she is interested in applying to. I am connected and powered up on my laptop; around me virtually everyone is either working on a laptop, talking on a mobile phone, or chatting with colleagues and friends. This is work in the modern day. We are all in the Digital Workplace and the upside for me is that Rose and I are together. When I log off we can chat about the day ahead. This must be a good thing for all of us.

And there is more to come outside the 'world of work.' The Digital Workplace will not only change the nature of work but will also change our societies – because it will change where we work, how we manage our close relationships and where we live. These are huge demographic and social issues, which will be important not only to managers but also to politicians, economists and social theorists – not to mention the architects and house builders.

Chapter 24

Predictions 2012–2013

So if that is the more distant future, what about the near term of this year and next? Here are my predictions for the key trends to watch out for.

Trend 1

The Digital Workplace will cause a wave of physical office redesign projects with real estate leading the shift.

Perhaps inevitably, the driving force in the Digital Workplace has so far been real estate reductions and the reshaping of office environments. The lead times in the physical world are far longer than in the digital, so organizations are trying to assess now what they will need on a physical work level five years from now. Will anyone come to an office? If so, who, when and why? The change in the Physical Workplace is being enabled by the rapid improvements in the Digital Workplace but lots of money is being wasted on offices that will be virtually empty in 2016.

Trend 2

The cultural impact, based on fears of isolation and fragmentation, will surface as a key human resource challenge and opportunity in the Digital Workplace.

From my own experience (and from what I hear from staff at IBM, who now have a decade of experience of working away from the office), the only downside to the portable nature of the Digital Workplace is a feeling of isolation from colleagues and the organization for which they work. The Digital Workplace enables, at its best, a consistent experience of work wherever you are, which is great for freedom and flexibility, but the HR challenge is to overcome the loss of the vital human connection that is necessary for productive work. Seeing colleagues from time to time, at least once a month perhaps, in different locations, makes a huge impact when set against having no physical contact at all for months on end.

Trend 3

The deplorable state of the usability of the Digital Workplace will start to be noticed as an obstacle to efficiency.

What is often ignored is that the digital world of work persists just as much when we are physically working in company-owned locations – offices, warehouses or plants – as it does when we are situated anywhere else. We will need to design Digital Workplaces that flex based on where we are; what we require in order to work successfully in an office is different from what we need for efficient work on a train, in a café or from our home office. Either way, Digital Workplaces need to offer a consistent and appropriate experience of work. Currently this is not the case, with the user experience of the fragmented, multiple-identity-requiring and chaotic digital worlds offered by most organizations, providing major challenges to those required to use them for work.

Yes, we love to have our work with us wherever we are, and we are mightily impressed by the speed and portability of today versus three years ago, but on a usability level the experience is still, more often than not, really quite poor. In a few years' time, what we tolerate today will come to seem as antiquated as 'dial up' does now.

Trend 4

The 'digital examples' to follow will work in the way Facebook, Google and Twitter are organized as companies – with very few people producing huge amounts of financial value and with major traditional corporates trying to reshape how work happens.

The current financial crisis is two-fold in my mind. Partly it is economic, due to debt, banking and liquidity – and this will eventually be resolved.

The larger crisis, though, may never be fully resolved with a new economic model emerging, where technology allows a new relationship between people and productivity – with fewer people required to produce higher financial values. If we look at the new technology giants such as Facebook and Google, they employ very small staffs and produce massive financial value. Such companies are models of a new economy where technology replaces people at frightening levels. Already, if you visit a modern factory today, you will see very small numbers of human beings in very large spaces. More traditional organizations will increasingly try to emulate the 'Google/Facebook model' as digital work drives down the cost of production. What will this mean for jobs? Who knows? But, entirely new industries will surely emerge out of this radical restructuring.

Trend 5

Governments will lead the drive at policy level for a fundamental shift to digital working and mobility, with organizations struggling to match the pace of change.

London hosts the Olympics in 2012 and at Government level there is a drive to promote flexible working for three weeks around the Games. Organizations are being required by the Government to change their policies because the Digital Workplace can take the strain, and these organizations will never look back once the

Olympics finish as habits will have been changed. In Holland, Finland, the US and UK, government policy loves the Digital Workplace – less traffic, less sickness, reduced carbon, fewer accidents on the roads, business as usual when bad weather strikes, happier home lives – and this top-down push will accelerate corporate-wide shifts in how and where work happens.

Trend 6

The Digital Workplace will grow and develop as a more general world of work and technology and not as a 'bigger, better intranet.'

Telephones, mobile devices, video and audio conferencing, micro-blogging, HR systems, email, customer social media and the wider range of work and technology all make up the Digital Workplace. Intranets will continue to be essential core services but the Digital Workplace is not a bigger, better intranet and never will be.

As an illustration of the point, let's take the example of transport. We will always have trains – better and faster no doubt – but the train will never be *all* transportation; both trains and transport are important in their respective roles but we must not confuse the two.

We will drive forward intranets in 2012 and also enrich Digital Workplaces, but let's make sure we know which is which.

Trend 7

Working across geography and time zones will increase and power the expansion of the Digital Workplace with new innovation and collaboration opportunities.

When you cast off the shackles of the physical world, you can fly. Activities, collaborations and projects are possible in a digital working world that are simply impossible in the physical space. Organizations will exploit the Digital Workplace to assemble new teams, projects and processes that span time zones and regions. This will show that the Digital Workplace is not just a way to work from home (the least imaginative use of the Digital Workplace really) but a means to innovate and collaborate in fresh, surprising ways, leading to new services, products and efficiencies.

Trend 8

'Bring Your Own Device' trends will drive the Digital Workplace towards mobile services accessed via single login details secured at the point of entry.

A senior manager I know tells the story of his grandfather, who was given work clothes when he began work in Italy a century ago; this manager's father was given a driving licence when he began work; and the manager himself was given a PC.

What will new hires get in the future when they join? Probably nothing but a secure identity and login to a set of Cloud-based services. People will use their existing tablets, phones, laptops and be happy to just 'hook into the corporate system' from them. 'Bring Your Own Device' (BYOD) will become the new normal and this gradual trend will just tick along in 2012. The question is when they log on, how good is the quality of what they access?

Trend 9

The Digital Workplace field will create increased anxiety and risk management concerns at senior levels, leading to more strategic controls.

Big change always comes with problems and, aside from isolation mentioned earlier, the other huge obstacle is security and risk management. If people are increasingly 'anywhere', how can the work they do remain secure and not expose the organization to unknown risks and legal dangers? The large technology firms are ploughing investment into security in the Digital Workplace but anxiety levels at the CEO and other C levels will rise as problems surface and gain attention in the media. Remember, getting these things wrong can land a CEO in prison – that danger focuses minds.

Trend 10

The Digital Workplace will start to be regarded as a major business opportunity rather than simply a replacement for physical offices.

The Digital Workplace is currently seen as a cheaper, more flexible way to work than Physical Workplaces, something of a replacement. What will start to develop is a belief that the Digital Workplace is not only a huge area of business in its own right, with new B2B services and sectors, but also that it offers a better, more productive and innovative space in which to work than physical offices. It will take on a shape and stature of its own and this journey into work/technology – that really began with the telephone – will become an ever richer, more diverse and potent place in which to do business.

References

1. http://www.whoinventedit.net/who-invented-the-telegraph.html
2. http://en.wikipedia.org/wiki/Alexander_Graham_Bell
3. Charles Dickens, *Hard Times*, 1854
4. http://www.flexibility.co.uk/flexwork/location/homeworking-statistics-2009.htm
5. http://www.flexibility.co.uk/issues/WLB/long-hours-health.htm
6. http://www.teleworkresearchnetwork.com/research/people-telecommute
7. http://mobithinking.com/mobile-marketing-tools/latest-mobile-stats
8. http://www.knowledgeatwharton.com.cn/index.cfm?fa=viewfeature&articleid=1549&languageid=1
9. http://www.melcrum.com/articles/pitney_bowes_CEO.shtml
10. www.mikecritelli.com
11. http://www.pb.com/Our-Company/Corporate-Responsibility/Our-People/Engagement-and-Development.shtml
12. http://www.watsonwyatt.com/research/printable.asp?id=W-488
13. http://www.bestcompanies.co.uk
14. http://www.thefutureofwork.net/assets/Citrix_EMEA_Managing_Pple_Cant_See.pdf
15. http://www.regus.presscentre.com/Press-Releases/REGUS-AND-UNWIRED-LAUNCH-GLOBAL-WORKPLACE-REPORT-2242.aspx

16. Arup, Foresight Report: Living Workplace, 2011; http://www.driversofchange.com/make/research/livingworkplace

17. Case study based on transcript of interview between Hal Stern and Ann Bamesberger, original Sun Microsystems website

18. http://www.smart2020.org/_assets/images/Open_Work_Energy_WP_02-25-09.pdf

19. http://tedxtalks.ted.com/video/TEDxManhattanBeach-Thomas-Suare

20. http://www.cbs.nl/en-GB/menu/themas/financiele-zakelijke-diensten/publicaties/artikelen/archief/2011/2011-3452-wm.htm

21. www.odesk.com

22. Original interview with Gary Swart, February 2011

23. 'Flexible working hours at the BMW Group' brochure, BMW Group 2002 at: http://www.bmwgroup.com/e/0_0_www_bmwgroup_com/unternehmen/publikationen/aktuelles_lexikon/_pdf/FlexWork_E.pdf

24. Interview with Marc Groenninger, Head of Human Resources Systems, BMW

25. http://www.telecompaper.com/news/bmw-implements-twist-teleworking-project

26. Cisco Connected World Technology Report, 2011: http://www.cisco.com/en/US/netsol/ns1120/index.html

27. Kate Lister and Tom Hamish, The State of Telework in the US, 2011: http://www.workshifting.com/downloads/downloads/Telework-Trends-US.pdf

28. http://urwebsrv.rutgers.edu/medrel/viewArticle.html?ArticleID=5284

29. http://www.usyd.edu.au/news/84.html?newsstoryid=2262

30. http://news.bbc.co.uk/1/hi/uk/4471607.stm

31. http://seeit.mit.edu/Publications/CrackBerrys.pdf

32. Stephen R. Covey, *The 7 Habits of Highly Successful People*, first published 1989

33. www.teleworkresearchnetwork.com

34. Sherry Turkle, *Alone Together*, 2011

35. Jaron Lanier, *You Are Not A Gadget: A Manifesto*, 2011

36. http://www.bbc.co.uk/news/technology-16055310

37. Mark Dixon and Philip Ross, *Work: Measuring the Benefits of Agility at Work*, 2011: http://www.regus.presscentre.com/imagelibrary/downloadMedia.ashx?MediaDetailsID=26066

38. http://www.fmwf.com/media-type/news/2010/02/working-from-home-is-good-for-employee-engagement-and-good-for-business

39. http://gorowe.com/know-rowe/what-is-rowe

40. http://www.thisislondon.co.uk/lifestyle/article-23749476-rise-of-the-results-oriented-work-environment.do

41. Cali Ressler and Jody Thompson, *Why Work Sucks and How to Fix It: The Results-Only Revolution*, 2008

42. Daniel H. Pink, *Drive: The Surprising Truth About What Motivates Us*, 2011

43. http://www.businessweek.com/magazine/content/06_50/b4013001.htm

44. http://gorowe.com/wordpress/wp-content/uploads/2009/12/why_work_sucks_intro_and_chapter_1.pdf

45. http://www.americanprogress.org/issues/2010/11/rowe111010.html

46. http://www.microsoft.com/presspass/emea/presscentre/pressreleases/JPC_NWoW_25112010.mspx

47. Eric van Heck, *New Ways of Working – Microsoft's Mobility Office*, RSM Insight 02 Paper, Q1 2010

48. http://www.slideshare.net/alexada/white-papera-new-way-of-working-microsoft-netherlands-external

49. http://www.slideshare.net/vanwilgenburgh/new-world-of-work-microsoft-netherlands-case-study

50. http://www.retail-week.com/vodafone-launches-staff-mobile-intranet/1333024.article

51. http://mashable.com/2010/12/13/idc-mobile-apps-study

52. http://www.businessweek.com/technology/content/nov2010/tc2010111_326784.htm

53. http://www.ekahau.com/products/products-overview.html

54. http://en.wikipedia.org/wiki/Augmented_reality

55. http://www.layar.com/blog/2011/03/16/video-layar-3d-augmented-office

56. https://www.yammer.com/about/about

57. http://www.headshift.com

58. http://www.bbc.co.uk/news/uk-england-manchester-12308600

59. http://massively.joystiq.com/2009/11/04/linden-lab-launches-second-life-enterprise-beta-second-life-wor

60. http://www.hypergridbusiness.com/2010/08/second-life-discontinues-enterprise-platform

61. http://www-03.ibm.com/press/us/en/pressrelease/27831.wss

62. http://labs.oracle.com/projects/mc/mpk20.html
63. http://www.cisco.com/web/about/ac227/csr2010/report-card/index.html
64. http://newsroom.cisco.com/dlls/2010/ts_012610.html
65. http://home.cisco.com/en-us/telepresence/umi
66. http://researcher.ibm.com/view_project.php?id=1231
67. http://www.intranetjournal.com/articles/200209/ij_09_25_02a.html
68. http://www.prescientdigital.com/articles/intranet-articles/employee-social-networking-case-study
69. http://www.steptwo.com.au/papers/kmc_gametheory/index.html
70. http://www.jmorganmarketing.com/wp-content/uploads/2010/01/ZDNet-Zappos-SR.pdf
71. http://about.zappos.com/our-unique-culture/zappos-core-values/build-positive-team-and-family-spirit
72. http://www.lovemachineinc.com/2011/03/great-kevin-rose-interview-including-our-new-office
73. http://trial.sendlove.us/trial/index.php
74. http://www.lovemachineinc.com/2011/02/a-different-model-for-building-software
75. http://www.lovemachineinc.com
76. http://www.brighthub.com/environment/green-computing/articles/33216.aspx
77. http://www.tuc.org.uk/social/tuc-18791-f0.cfm
78. http://www.gartner.com/it/page.jsp?id=876512
79. http://en.wikipedia.org/wiki/Remote_surgery
80. http://news.bbc.co.uk/1/hi/sci/tech/1552211.stm

81. http://images.businessweek.com/ss/10/06/0602_ceo_guide/4.htm

82. http://www.intouchhealth.com/products_rp-7_robots.html

83. http://www.aethon.com/spotlight/case_studies.php

84. http://www.businessweek.com/technology/content/jun2010/tc2010061_798891.htm

85. http://www.post-gazette.com/pg/06167/698696-96.stm

86. http://www.gohyper.com/products/hyperactive-bob

87. http://www.thisislondon.co.uk/restaurants/review-23553470-dining-goes-digital-at-inamo.do

88. http://www.headthere.com/products.html

89. http://online.wsj.com/article/SB10001424052748704073804576023613748146944.html#ixzz18gNNZ3e2

90. http://www.cloudsmagazine.com/miscellaneous/tmsuk-4-a-robot-that-makes-the-purchase.html

91. http://www.engadget.com/2009/01/23/tmsuk-t-34-robot-speaks-softly-carries-a-big-net

92. Abraham Maslow, *A Theory of Human Motivation*, 1943

Book list

Jim Collins, *Good to Great: Why Some Companies Make the Leap… and Others Don't*, 2001. HarperBusiness.

Stephen R. Covey, *The 7 Habits of Highly Successful People*, first published 1989.

Jeff Jarvis, *What Would Google Do?* 2009. HarperBusiness.

Jaron Lanier, *You Are Not a Gadget: A Manifesto*, 2011. Allen Lane.

Daniel H. Pink, *Drive, The Surprising Truth About What Motivates Us*, 2011. Canongate Books.

Cali Ressler and Jody Thompson, *Why Work Sucks and How to Fix It: The Results-Only Revolution*, 2008. Portfolio.

Sherry Turkle, *Alone Together*, 2011. Basic Books.

About Digital Workplace Group

The Digital Workplace Group provides research, interaction, measurement and consultancy across the field of work and technology in major organizations globally. There are several businesses within the Group, including the Intranet Benchmarking Forum, the Digital Workplace Forum and Digital Workplace Consulting.

Digital Workplace Group
www.digitalworkplacegroup.com

Intranet Benchmarking Forum
www.ibforum.com

Digital Workplace Forum
www.dwforum.com

Digital Workplace book site
www.digitalworkplacebook.com

Email
info@digitalworkplacegroup.com
info@ibforum.com
info@dwforum.com

Acknowledgements

Special thanks go to Steve Bynghall who worked as the Content Producer, contributing much of the case study content of the book. Also big thanks to Alison Chapman who acted with great patience and attention as an Editor throughout the process of turning an idea into the finished product. I must also thank Toast Design for the superb work in making the design of the book gel so well with the style of content. Huge thanks also to the expert team across the Digital Workplace Group and particularly those within the Intranet Benchmarking Forum, including Helen Day, Elizabeth Marsh, Angela Pohl and Mark Silverman. Special mention for Nancy Goebel for her work with the marketing and commercializing of the book and to Paul Levy for challenging the ideas in it and motivating me to take the extra step. There is also a list of others in no particular order who helped me with the book and whom I would like to thank, including Andrew Marr, Sam Marshall, Chris Tubb, Cheryl Lesser, John Baptista, Nic Price, Ross Chestney, David Smith, Jeff Ramos, Tim Pallant, Sue Frampton and to my family and friends and all the many great people across so many different organizations who encouraged me with the work.

Lightning Source UK Ltd.
Milton Keynes UK
UKOW06f0758281116
288701UK00021B/787/P

9 781457 510960